NEW TRENDS IN OPTICAL NETWORK DESIGN AND MODELING

IFIP - The International Federation for Information Processing

IFIP was founded in 1960 under the auspices of UNESCO, following the First World Computer Congress held in Paris the previous year. An umbrella organization for societies working in information processing, IFIP's aim is two-fold: to support information processing within its member countries and to encourage technology transfer to developing nations. As its mission statement clearly states,

IFIP's mission is to be the leading, truly international, apolitical organization which encourages and assists in the development, exploitation and application of information technology for the benefit of all people.

IFIP is a non-profitmaking organization, run almost solely by 2500 volunteers. It operates through a number of technical committees, which organize events and publications. IFIP's events range from an international congress to local seminars, but the most important are:

- The IFIP World Computer Congress, held every second year;
- open conferences;
- working conferences.

The flagship event is the IFIP World Computer Congress, at which both invited and contributed papers are presented. Contributed papers are rigorously refereed and the rejection rate is high.

As with the Congress, participation in the open conferences is open to all and papers may be invited or submitted. Again, submitted papers are stringently refereed.

The working conferences are structured differently. They are usually run by a working group and attendance is small and by invitation only. Their purpose is to create an atmosphere conducive to innovation and development. Refereeing is less rigorous and papers are subjected to extensive group discussion.

Publications arising from IFIP events vary. The papers presented at the IFIP World Computer Congress and at open conferences are published as conference proceedings, while the results of the working conferences are often published as collections of selected and edited papers.

Any national society whose primary activity is in information may apply to become a full member of IFIP, although full membership is restricted to one society per country. Full members are entitled to vote at the annual General Assembly, National societies preferring a less committed involvement may apply for associate or corresponding membership. Associate members enjoy the same benefits as full members, but without voting rights. Corresponding members are not represented in IFIP bodies. Affiliated membership is open to non-national societies, and individual and honorary membership schemes are also offered.

NEW TRENDS IN OPTICAL NETWORK DESIGN AND MODELING

IFIP TC6 Fourth Working Conference on Optical Network Design and Modeling February 7–8, 2000, Athens, Greece

Edited by

Alexandros A. Stavdas
National Technical University of Athens
Greece

KLUWER ACADEMIC PUBLISHERS
BOSTON / DORDRECHT / LONDON

Distributors for North, Central and South America:
Kluwer Academic Publishers
101 Philip Drive
Assinippi Park
Norwell, Massachusetts 02061 USA
Telephone (781) 871-6600
Fax (781) 871-6528
E-Mail <kluwer@wkap.com>

Distributors for all other countries:
Kluwer Academic Publishers Group
Distribution Centre
Post Office Box 322
3300 AH Dordrecht, THE NETHERLANDS
Telephone 31 78 6392 392
Fax 31 78 6546 474
E-Mail <services@wkap.nl>

Electronic Services <http://www.wkap.nl>

Library of Congress Cataloging-in-Publication Data

IFIP TC6 Working Conference on Optical Network Design and Modeling (4th: 2000: Athens, Greece)
New trends in optical network design and modeling: IFIP TC6 Fourth Working conference on Optical Network Design and Modeling, February 7-8, 2000, Athens, Greece / edited by Alexandros A. Stavdas.
p. cm. — (International Federation for Information Processing ; 62)
Includes bibliographical references and index.
ISBN 0-7923-7355-3 (alk. paper)
1. Optical communications—Congresses. 2. Fiber optics— Congresses. 3. Computer networks—Congresses. 4. Optoelectronics—Congresses. I. Stavdas, Alexandros A. II. Title. III. International Federation for Information Processing (Series) ; 62.

TK5103.59.I35 2001
621.382'7—dc21 2001022475

Printed on acid-free paper.

Printed in the United States of America.

Contents

Foreword

This is a very exciting time for working in optical networks. The deployment of WDM systems continues at an accelerated pace thanks to the low cost and high reliability of optical components. This primarily refers to optical amplifiers, fixed/tunable filters followed and recently true optical cross-connects. The demand for capacity, driven by the Internet explosion, is higher than ever with no signs of saturation. The emergence of data-centric networking calls for networks with fewer and more robust layers, enhanced control and an integrated management platform. Networks with fewer optical elements are sought. We witness the co-existence of a multitude of transport mechanisms (SDH, IP, ATM) together with a diversification in the offered services with widely varying QoS requirements. At the same time, components and subsystems for very high speed optical networks are maturing opening up a new dimension and offering even more design options.

After Vienna (1997), Rome (1998) and Paris (1999), the Fourth Working Conference on Optical Network Design and Modeling was held in Athens, Greece February 7th-8th, 2000. The aim was to present the most recent progress in terms of network architectures, design, operation and maintenance, and planning. The book you have in your hands covers the topics raised in this Working Conference. We sincerely hope the reader would not only find this book informing but also stimulating in generating new ideas.

Alexandros Stavdas
Department of Electrical and Computer Engineering
National Technical University of Athens

Acknowledgments

The Editor of this book is extremely grateful to the authors of the articles to be found in the subsequent pages of this book and to Mrs. Admela Jukan for her participation in the preparation of Conference. The Editor would also like to acknowledge the significant contribution of Mr. Charalambos Skoufis, PhD candidate in our University, in the preparation of this Conference and for taking good care of the final version of the manuscripts.

Multi-Granularity Optical Networks

Multiplexing and Switching connections in new dimensions

C. Blaizot, E. Dotaro, L. Noirie, A. Jourdan
Alcatel, Corporate Research Center, Optical System Department,
Route de Nozay 91461 Marcoussis Cedex

Key words: WDM, Band, multiplexing, MG-OXC, granularity, OXC, grooming

Abstract: Considering the telecommunication traffic boom, one can wonder if routing nodes will provide the corresponding switching capacity. Introducing Multi-Granularity Optical Networks concept, this paper gives a cost-effective solution. First, a MG-OXC architecture is proposed, then the "natural" distribution of the traffic among the granularities is presented. Finally, a Multi-Granularity planning process is described. Based on the most advantageous feature of optical technologies, the proposed approach solves the switching matrix size limitation.

1. INTRODUCTION

Nowadays, we are witnessing a new telecommunication traffic boom, for which very high transmission and switching capacities are required. Tbit/s transmission capacity can be achieved thanks to optical fibres. The number of transmitted wavelengths in one fibre which has recently been demonstrated in research laboratories exceeds 100 wavelengths per fibre and even more for terrestrial transmission (for example 150x10 Gbit/s over 400 km, see [1]) and even for submarine transmission (for example 100x10.7 Gbit/s, see [2]). Some suppliers announce commercial systems with 80, 160 and even 256 wavelengths per fibre. In optical networks, the optical cross-connects (OXC) will have many fibres, each one carrying many wavelengths. The switching of such a high capacity raises severe scalability issues for OXC design.

In this paper, we first describe a Multi-Granularity Optical Cross-Connect (MG-OXC) solution to switch all this high-capacity traffic. This solution is based on wavelength, band and fibre switching and squeezes the most out of optical technologies. Then we study what can be the distribution of the traffic between wavelength switching, band switching and fibre switching on the basis of a typical meshed network. Finally, a planning methodology is described: it optimises the grooming to avoid too much multihopping between the different granularities through the optical backbone network, which decreases the required number of optical ports and the OXC switching matrix size, and thus the cost of the OXC-based network.

2. MULTI-GRANULARITY OPTICAL CROSS-CONNECT

2.1 Required capacity for wavelength-switching cross-connect

For very high capacity backbone networks (may be up to Pbit/s for future networks), with large number of fibres with over 100 wavelengths carried by each fibres, the optical cross-connects (OXC) will have many input/output fibre ports, each one carrying many wavelengths. An OXC may have for example 100 input fibres with 160 wavelengths each, each wavelength carrying 10 Gbit/s data. The total throughput to be switched by such a node is 160 Tbit/s. The total number of input wavelengths is N=16000. If one wants to switch all the wavelengths individually, the require space-switching matrix is 16000x16000. A mono-block switch with this size is not achievable because of the high number of cross-connection points: 256.10^6 (N^2). A Clos architecture may be used, but this requires a first stage with 180 "90x180" switches, a second stage with 180 "180x180" switches and a third stage with 180 "180x90" switches. The total number of cross-connection points ($4N\sqrt{2N}$ for Clos architecture) is still high: about 12.10^6. Space-switch matrices with such high number of cross-connection points cannot be envisaged because of the high number of required components and because of the complexity of the electrical control of such a matrix.

2.2 Solutions to decrease the port number

To be able to switch all this traffic (160 Tbit/s), one should group this traffic to higher granularity to decrease the number N of switched entities.

2.2.1 To increase the capacity of each wavelength

One first solution is to carry higher bit-rate on each wavelength like in SONET/SDH networks: the granularity can go from OC-3/STM-1 (155 Mbit/s) to OC-12/STM-4 (622 Mbit/s), OC-48/STM-16 (2.5 Gbit/s) and OC-192/STM-64 (10 Gbit/s) today. Transmission of 40 wavelengths with 40Gbit/s (future OC-768/STM-256) carried by each wavelength has already been achieved ([3]). To increase the wavelength bit-rate impacts the transmission. Issues encountered are chromatic dispersion, non-linear effects, polarisation mode dispersion: the higher the per-wavelength bit-rate is, the more detrimental these transmission effects are. Moreover simple transmitters and receivers at such high bit-rate are not yet available.

2.2.2 To group the wavelengths

Another solution is to keep the 10 Gbit/s bit-rate for each wavelength, and to group the wavelengths to bands to switch them together. For our example one band can be composed of 16 wavelengths, each fibre carrying 10 bands. The total number of bands to switch is then N=1000 and the total throughput of the OXC is still 160 Tbit/s.

The required switching matrix is then 1000x1000. A mono-block switch with this size may be achievable (see [4] for 576x576 matrix), the number of cross-connection points is 10^6 (N^2). A Clos architecture can be used and requires a first stage with 45 "23x45" switches, a second stage with 45 "45x45" switches and a third stage with 45 "45x23" switches. The total number of cross-connection points ($4N\sqrt{2N}$) is quite low: about 220.10^3. Optical space-switching matrices with such a number of cross-connection points have already been envisaged [5].

2.3 Multi-Granularity OXC architecture

With a lot of fibres at each OXC, the band granularity may sometimes be too small, and the switching may be done at the fibre level. More generally, one may want to switch a part of the total traffic at the fibre level, another part at the band level and a last part at the wavelength level. What we need in this case is a Multi-Granularity Optical Cross-Connect (MG-OXC). The figure 1 shows a possible architecture for such a MG-OXC.

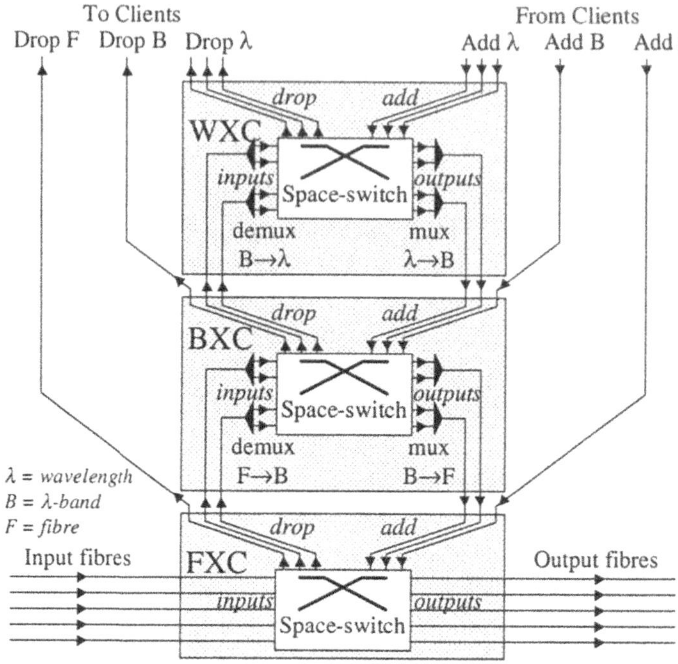

Figure 1. Multi-Granularity Optical Cross-Connect (MG-OXC)

The optical signals are first switched at the fibre level, in the Fibre Cross-Connect (FXC): some fibres are directly added/dropped towards the clients (they may be demultiplexed to wavelengths for clients, but the demultiplexers are not represented on the figure), others are directly switched to output fibres, few ones are added/dropped to the band level. These add (resp. drop) fibre ports of the FXC are connected to the output (resp. input) band ports of the Band-Cross-Connect (BXC) through band-to-fibre multiplexers (resp. fibre-to-band demultiplexers). Some input bands of the BXC are directly added/dropped towards the clients, others bands are directly switched to output ports of the BXC, and few ones are added/dropped to the wavelength level. These add (resp. drop) band ports of the BXC are connected to the output (resp. input) wavelength ports of the Wavelength-Cross-Connect (WXC) through wavelength-to-band multiplexers (resp. band-to-wavelength demultiplexers). Some input wavelengths of the WXC are added/dropped towards the clients, others wavelengths are switched to output ports of the WXC.

On the figure, we did not represent optical amplification which will be needed inside the OXC and at the input and output fibres. 3-R regeneration may also be required at the input and/or output fibres. Wavelength (resp.

band) conversion will also be needed at the output ports of the WXC (resp. BXC) but they have not been represented to simplify the drawing.

The figure 2 gives an example of what could be the distribution of the traffic in the MG-OXC, for 160 Tbit/s external throughput (100 input/output fibres, with 10 bands per fibre, 16 wavelengths per band, 10 Gbit/s data on each wavelength). We also have add/drop traffic (add/drop ports): 10 fibres (16 Tbit/s), 30 bands (4.8 Tbit/s) and 80 wavelengths (0.8 Tbit/s). The total throughput to be switched by the node is thus 181.6 Tbit/s. We will see later that the aggregate capacity of the switching matrices is higher because of internally generated traffic for grooming purpose.

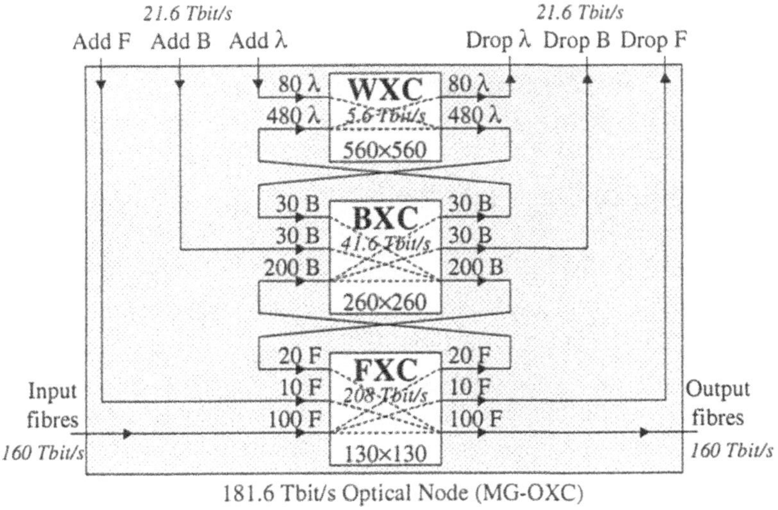

Figure 2. Example of traffic distribution in the MG-OXC

Among the 100 input fibres (we don't take into account the 10+20=30 add fibre ports of the FXC), 70 are cross-connected to the output fibres at the FXC level, 10 are directly dropped and 20 are dropped to the BXC input ports. The 10 add fibre-ports and the 20 fibres coming from the outputs of the BXC are cross-connected to the 30 remaining output fibres of the FXC. The 20 fibres dropped to the BXC represent 200 bands, among which 140 are cross-connected to the output ports of the BXC (grooming, to rearrange the bands inside some fibres), 30 are directly dropped and 30 are dropped to the WXC. These 30 bands correspond to 480 wavelengths, among which 400 are cross-connected to the output ports of the WXC (grooming at the wavelength level) and 80 are directly dropped. The 60 add ports of the BXC (30 add band-ports and the 30 bands coming from the output ports of the WXC) are cross-connected to the 60 remaining output ports of the BXC, and the 80 add wavelength-ports are cross-connected to the 80 remaining output ports of the WXC.

For this example, this will require a 130x130 space-switch matrix for the FXC (208 Tbit/s), a 260x260 space-switch matrix for the BXC (41.6 Tbit/s) and a 560x560 matrix for the WXC (5.6 Tbit/s). The aggregate capacity is about 255 Tbit/s, which is higher than the total throughput to be switched by the node (181.6 Tbit/s). This is due to the fact that for the presented architecture, the same signals are switched at different granularities, and some may go through one stage twice for grooming purpose. The total number of cross-connection points is only about 400.10^3 even without Clos architecture. With Clos architecture for each space switch, this number is then only about 110.10^3. With such a MG-OXC architecture, the complexity of the space-switch matrix is decreased by a factor of about 100 compared to an only wavelength switching matrix for the whole traffic, and we have still some flexibility to switch a part of the traffic at the wavelength level.

2.4 Required technologies

To implement this MG-OXC, some key technologies are required : space-switching, multiplexing/demultiplexing, wavelength or band translation and in-node amplification.

First, we need a technology to switch the wavelength (WXC), a technology to switch the band (BXC) and a technology to switch the fibres (FXC). One may envisage electronic switching matrix for wavelength switching (WXC), with opto-electronic interfaces. A lot of optical technologies are also available to do the wavelength-switching ([4-6] for several wavelength switching technologies). For the band (BXC) and fibre (FXC) switching, electronic cannot do it, only optical technologies can. Moreover, the same technology can switch wavelengths, bands and fibres, provided that the optical bandwidth is large enough to switch all the wavelengths (typically 1500nm~1620nm, or even 1300nm~1620nm). This ability to switch everything from a single wavelength to a whole fibre multiplex is interesting on the up-gradability point of view : the same matrix technology firstly switches wavelengths when the traffic through the OXC is low (~1 Tbit/s), it can switch bands when the traffic begins to increase (~10 Tbit/s), and it will be able to switch fibres later when the traffic becomes very high (~100 Tbit/s). Some possible technologies are the Micro-ElectroMechanical System technology (MEMS, see for example [5]), system based on index-matching injection (see for example [6]) or bulk optomechanical free-space switches (see [4] for instance).

At the BXC and WXC input/output, some (de)multiplexing devices are required. Wavelength (de)multiplexers are now current WDM devices. Some devices can (de)multiplex up to 100 wavelengths [7]. Their design can be adapted to the demultiplexing of bands of several wavelengths.

Wavelength converters are already existing devices. They can be opto-electronic devices (photodiode-receiver cascaded with classical transmitter working at the new wavelength) or all optical converters [8]. Band converters can be simply made with a demultiplexer cascaded with n (n=16 in our example) parallel wavelength converters followed by a multiplexer.

In-node optical amplification may use the same technology as the amplification for transmission, with some simplifications to have low-cost amplifiers because requirements may be relaxed inside the node.

3. "NATURAL" DISTRIBUTION OF THE TRAFFIC IN THE MG-OXC

3.1 From mono-block node to sub-nodes stack

On a network point of view, the 3 granularities are "naturally" detected, demonstrating their usefulness. Although the planning is only based on the wavelength granularity, we can observe that, in each OXC, a large amount of traffic can be processed at the band and fibre levels, as if it was a "natural" behaviour.

To quantify this, there is first to obtain an initial planning and second to detect, in each node, which part of the traffic can be processed in which sub-node. The network is planned without specification on the nodes. The traffic routing is based on a Shortest Path in distance algorithm. The resources allocation is done at the wavelength level. The wavelength assignment process keeps as far as possible the same wavelength along each path, so that the wavelength translations are minimised. Then, starting from this simple initial network planning, it is possible to focus on the nodes. Looking at each resulting switching matrix and, without changing anything to the routing and resources allocation, each node is divided in a stack of sub-nodes (figure 3). The traffic distribution among the sub-nodes results from the detection of traffic correlation in the node, correlation in term of band or fibre. For example, wavelengths with the same input and output fibres, without any wavelength translation, are detected as composing a fibre port of the FXC.

It has to be noticed that no preparation of the traffic to favour high granularities is done. To favour fibres or bands of wavelengths, a first solution could be to use a dedicated wavelength allocation process filling up first fibres, second bands before considering the wavelength level. A second solution could be to prepare the traffic starting from the routing, as described in section 4.

Two kinds of traffic are differentiated: the *internal* traffic and the *external* traffic. The internal traffic corresponds to the traffic between the sub-nodes while the external traffic is the traffic coming from / going to the node's neighbours or to the clients. This internal traffic increases the size of the sub-nodes and represents the flexibility of the global node. For example, the architecture figure 1 is more flexible than the architecture figure 3. Figure 3, all the fibres go through the FXC, thus, it is possible to select the fibres going to the other sub-nodes. The architecture figure 3 is based on dedicated fibres directly connected to the right sub-node: this is less flexible but also less port consuming.

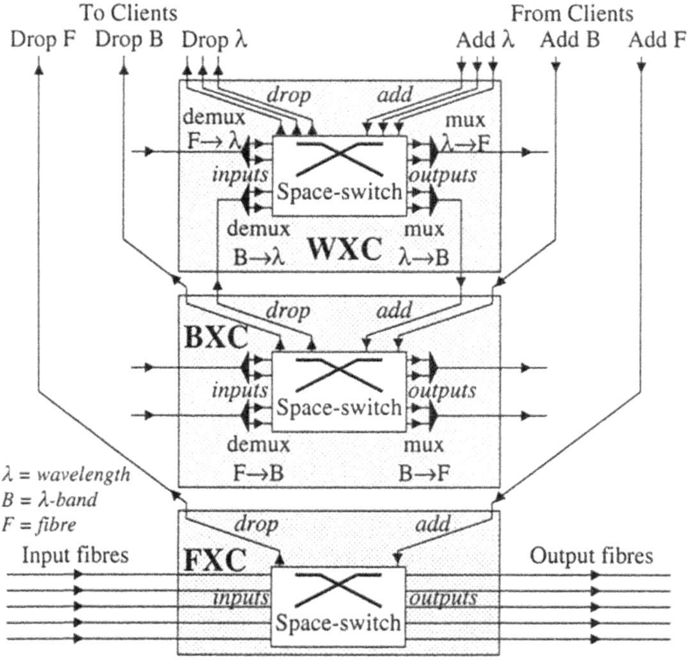

Figure 3. Sub-nodes piling up

3.2 Traffic distribution evaluation

To evaluate the amount of traffic processed by a sub-node, two parameters are identified: the *number of ports* and the *capacity*. The number of ports allows to compare the size of each sub-node. But, for example, a fibre port carries 32 wavelengths while a band port carries 8 wavelengths and a wavelength port only one. A port does not represent the same capacity. Considering the capacity means comparing the number of equivalent wavelengths switched by the sub-nodes. It allows to compare the amount of traffic treated in each sub-node. The difference is illustrated in the figure 4.

The capacity parameter has been chosen to study the part of traffic which could be switched at a granularity other than wavelength. More precisely, the traffic distribution is evaluated in capacity percentage, percentage calculated from an average on all the nodes of the network.

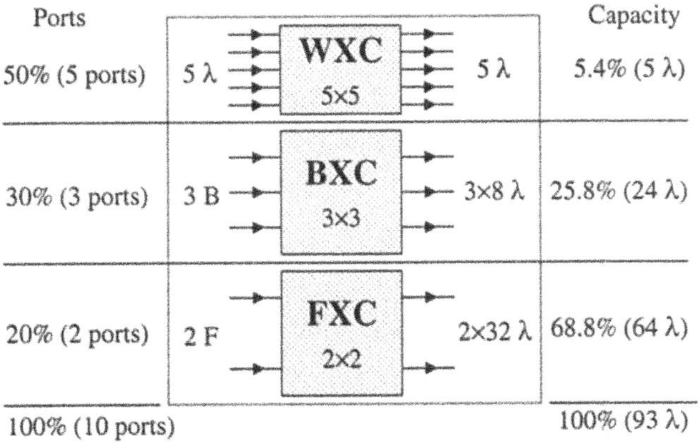

Figure 4. Comparison ports and capacity percentages

3.3 Traffic distribution results

The study has been done on the European optical network model COST239 [9] composed of 19 nodes (figure 5). The traffic matrix is complete, uniform or non-uniform according to the studied situation. A 32 wavelengths multiplex has been considered.

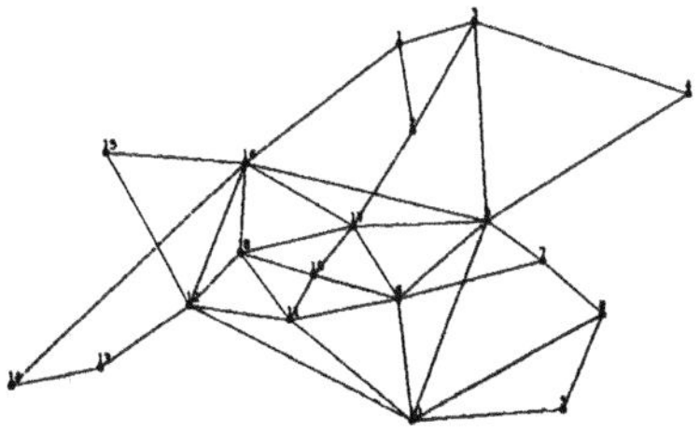

Figure 5. COST239 network

The following curves (figure 6) represent the percentage of capacity switched in the FXC, BXC and WXC for several traffic matrix multiplications (from one to 64) in several situations (uniform / non uniform traffic, with / without protection, 8 / 4 bands per fiber).

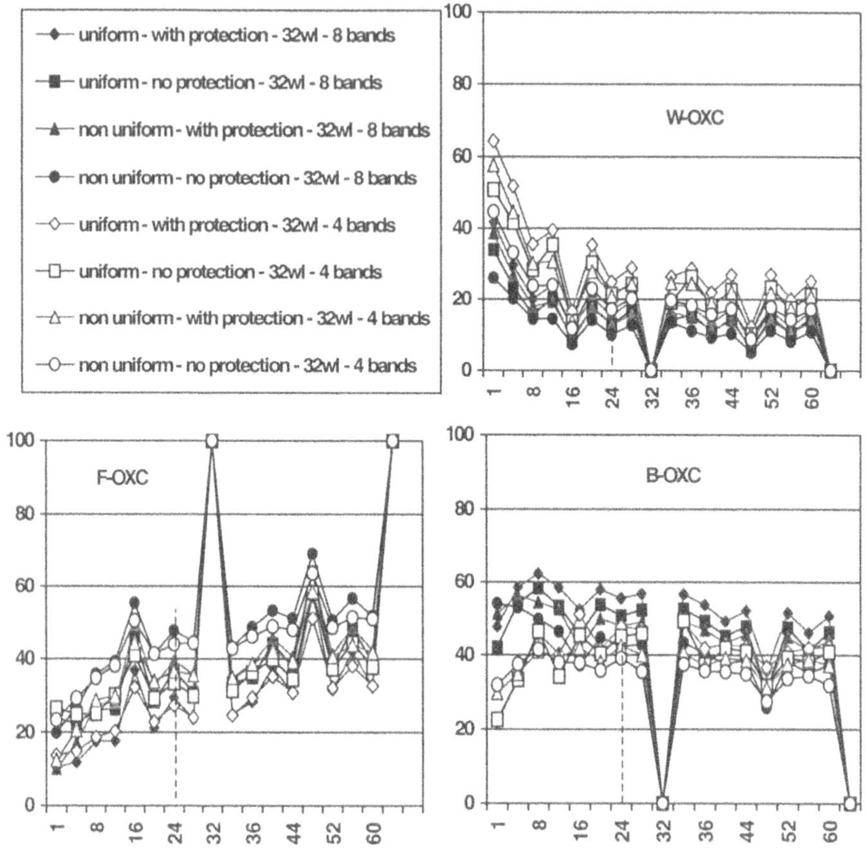

Figure 6. Percentage of capacity switched in the sub-nodes

For example, when the traffic demand is equal to 24 times the initial non uniform traffic matrix, without protection and 8 bands per fibre, the average distribution of the traffic in the sub-nodes is: 48% in FXC, 42% in BXC and 10% in WXC.

First visible result: 100% FXC, 0% BXC, 0% WXC when the traffic demand corresponds to **32** times the initial traffic matrix. This result is not surprising, and even expected. There are 32 wavelengths per fibre and each traffic demand between 2 nodes is a multiple of 32, in other words, the traffic demand between each pair of nodes corresponds to a demand in fibres.

Second visible result: when the traffic matrix is multiplied by Nx4 (respectively Nx8) and with a number of wavelengths per band equal to 4 (respectively 8), we expect to obtain only BXC because the traffic corresponds to a demand in band. It is easy to understand that FXC are obtained when several bands have the same direction. But what about the WXC? 40% to 15% of the capacity is still switched in the WXC. The main cause is the wavelength-granularity based planning process: the wavelength allocation process does not impose the same wavelength conversion, neither in the same node nor with the same wavelength shifting, to connections with the same source and destination. Thus, one wavelength conversion can occur in a band leading to the de-multiplexing in the WXC sub-node. In other words, the results obtained here are **lower than what we could obtain for the FXC and BXC** and **can be improved with an adapted planning process** taking into account the fibre and band granularities.

Third visible result: the shape of the curves are similar for all the situations. They are following almost the same behaviour, with the same peaks and troughs. We can also observe a kind of **pattern** on a range corresponding to the number of wavelengths per fibre with a low slope (around +0.3 for FXC, -0.1 for BXC and WXC). We can conclude that the traffic (uniform / non uniform, protection / no protection) and the band size have only an impact on the offset of the curve. This **offset** corresponds to a **gap in the multiplication coefficient of the traffic matrix**.

Fourth visible result: in this network, with 32 wavelengths per fibre, the average distribution of the capacity is rapidly under 40% and mainly **15% in the WXC** (and can be reduced), more than 20% and mainly **40% in FXC** and **45% in BXC.** To go near to the distribution presented with the figure 2 (around 3% WXC, 16% BXC, 81% FXC), there is to shift around 80% of the capacity to a bigger granularity (WXC to BXC, BXC to FXC) and /or to have a huge amount of traffic. Both hypotheses are realistic.

Now, changing the number of wavelengths per fibre (64 and 128 wavelengths), the following curves are obtained (figure 7). The situation is without protection, non uniform traffic. Apart from some peaks or troughs, mainly due to a traffic demand corresponding to a demand of fibres as explained in the first visible result, the **distribution is in line with the previous curves** mainly for the FXC.

The BXC offset observed for the situations "128 wl – 16 bands" and "64 wl – 8 bands" comes from the small size of the bands compared to the size of the multiplex. More traffic is required to fill up the complete multiplex, so more traffic is remaining in the BXC. In the same way, the WXC offset observed for the situations "128 wl – 4 bands" and "64 wl – 4 bands" comes from the big size of the bands (16 and 32 wavelengths). More traffic is required to fill up the bands, so more traffic is remaining in the WXC. Small bands look interesting, but it means more band ports than with big bands, obviously.

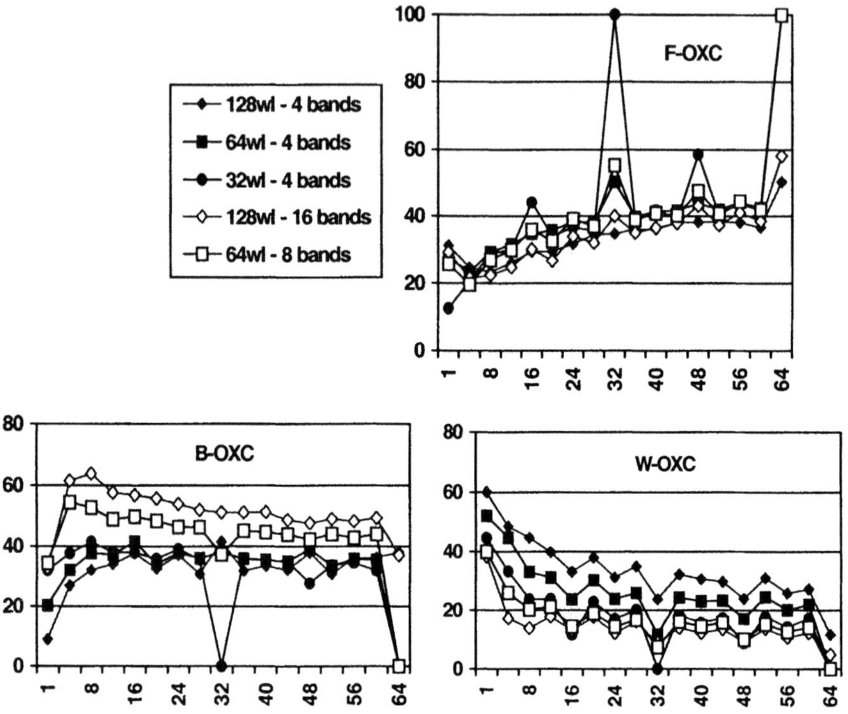

Figure 7. increase of the multiplex size

The behaviour of the **capacity distribution in the sub-nodes is stable**, only the offset of the distribution is changing according to the amount of traffic and the multiplex composition (number of wavelengths / bands). With an adapted planning process, we expect to improve the results. **Today**, for the COST239 network, without preparation of the traffic to obtain bands and fibres, and without considering additional capacity for flexibility, we obtain on average **15% WXC, 45% BXC, 40% FXC**. For tomorrow, the goal is to decrease the amount of traffic processed in WXC by favouring FXC and BXC as described in the following section.

4. PLANNING PROCESS ASPECTS

4.1 A new framework for optical network planning

The proposed scheme is based on the opportunity to switch various granularities in the optical domain. It represents a new step for the WDM deployment trends, overtaking previous visions [10-12]. The principle driver for the emergence of new WDM technologies is still derived from the increasing demand in terms of capacity. Before use of future combination with high-capacity packet switching (optical or not), the Multi-Granularities optical scheme remains a circuit-oriented approach with additional layers available. Therefore our goal is to define an exhaustive planning methodology where the whole range of circuit-granularity could be used at the right place with the right size. The evolution of optical node size and functionality has to face the evolution of demand, mainly in terms of amount of traffic. Our objective is to propose a flexible solution for the necessary migration of transport networks. While combining the very high capacity available in the optical domain and the flexibility given by the integration of Multi-granularity layers, we intend to squeeze the most out of optical technology. Many steps in this new approach have generic aspects, but in the following we focus on a dedicated planning method starting from an existing SDH/SONET layer which is at the present time the first potential client for a Multi-Granularity core optical layer.

4.2 A dedicated process is required

In the previous sections, we have shown that technological hardware solutions exist and can be available in short terms. Moreover, the analysis of planning according to typical topology and traffic evolution demonstrates that the architecture proposed is quite relevant for future transport network services. The concrete deployment of Multi-Granularity nodes within large optical core networks requires some associated planning methods in addition to the already well-known complexity of design [13].

A planning methodology is a complex problem including a lot of parameters, constraints and objectives. The introduction of new layers corresponding to new optical granularities has to be taken into consideration with care, otherwise the gain expected could be negligible. To be cost effective, an appropriate method has to be defined and has to reduce some unenthusiastic impacts of multihopping. The problem addressed here is the redundancy of resources in several layers that process the same traffic. From the switching point of view, it is cost-effective to switch in the same time numerous small-granularity connections multiplexed in a same high-granularity entity. In spite of this theoretical advantage, a bad multiplexing strategy may imply to de-multiplex a high-granularity connection in order to switch only few lower-granularity connections at another level. In this case, too many equipment are needed to process the same traffic.

The multihopping problem has an impact on the planning method but also on the migration strategy. If the network growth is only seen from a transmission point of view, an upgraded link is installed (e.g. WDM) according to the increase of traffic with an equivalent switching capacity at the source and sink nodes. In this situation, if the implementation of resources, switching this capacity, is done at a lower granularity level (e.g. large SDH nodes or wavelength cross-connect,...) then the introduction of high granularity optical layers would handle most of the capacity, making the previous resources redundant and useless.

Finally, if we want to optimize the benefits from the introduction of new granularities in the transport network, we have to define a dedicated process focused on multiplexing issues.

4.3 Main Objectives, Principles and Overall Process Description

Our main objective is to define a methodology for a cost effective planning, taking into account the new granularities within the optical layer. We want to propose relevant solutions facing the demand of large amount of traffic but also flexibility and upgradability considerations.

The central problem is to multiplex small-granularity connections in higher level connections. Hence more traffic is switched at each node using less resources and allowing to reduce the node size. From the planning point of view, we have to define an efficient way to arrange the distribution of connections on the topology in order to manage the correlation between connections.

As mentioned before, our process is starting from an existing SDH/SONET layer. In a first step a partitioning is done which gives the localization of WDM nodes and associated areas. The intermediate objective is to aggregate enough traffic coming from the SDH/SONET layer and to minimize the number of WDM nodes needed for the transport network. Although, as a result, the partitioning create some interdependence between connections starting from the same area and ending in another area. In this step the whole WDM topology is identified (nodes, links) and the traffic demand is established for the optical layer. The next step is to route the connections inside both layers, electrical and optical. At that time, the multiplexing operations begin with the TDM grooming. The goal in this step is to group together connections that have similar destiny in the network. In fact we try to generate wavelength connection with as less as possible demultiplexing that imply a switching at the electrical layer.

In the same way we try to find the best arrangement of wavelengths in the network to groom wavelengths into bands with the least band-to-wavelength demultiplexing. The last multiplexing operation, still based on the same logical rules, is to groom the bands into fibers. Finally a phase of resource assignment is needed to have the complete network planning.

The process is as flexible as possible flexible, with the possibility to modify the inputs and functionality at each step. For instance when a cost function is used, it could be replaced by real inputs or another cost function for a specific case.

4.4 Process description

4.4.1 Partitioning & Backbone topology definition

The partitioning problem is a traffic-based problem associated with topology and cost constraints. In order to decide if an initial SDH/SONET node should become a WDM node, first we have to evaluate the traffic that will go through this potential future WDM-node.

This traffic can be divided into 3 types:

- **ADD/DROP at the node**
- **ADD/DROP at nodes belonging to potential areas around the node**. It means that the node will catch the traffic of the area if it becomes a WDM-node.
- **Transit traffic at the node**. According to the node localization in the topology and the routing algorithm, the node will see more or less transit traffic. The amount of transit traffic has a major impact on the node size and has to participate in the decision whether or not the node should become a WDM-node.

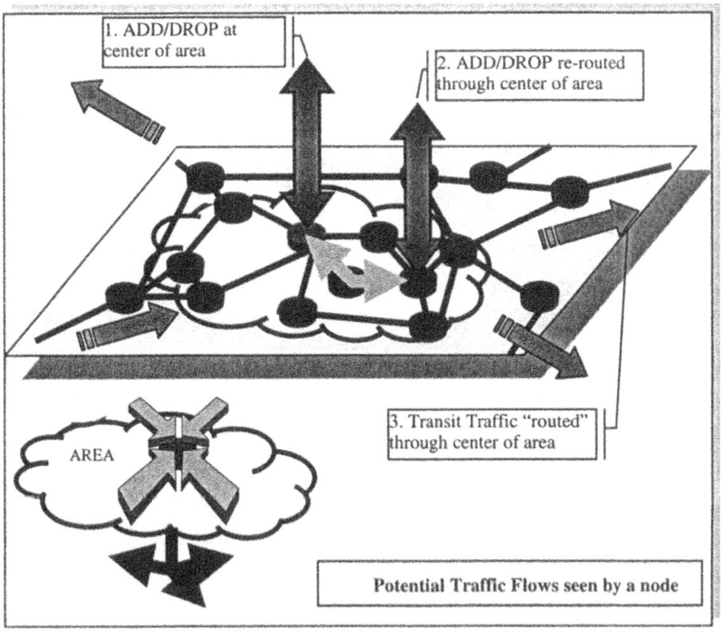

Figure 8. Potential traffic aggregation

The final decision of area creation has to be cost-based. It is a difficult problem because we have to compare the merits of different possible areas according to criteria which are future consequence of the global partitioning. This cost comparison is mainly based on the ratio of multiplexing factor between the two layers, but takes into account the increase of the path length. Hence we can evaluate an approximate gain for each potential area and decide the localization of the WDM nodes.

The gain function is first an evaluation of the number of ports needed to transport the connections starting from an area with or without the optical layer. Then the cost ratio between SDH/SONET and WDM ports and the cost ratio between transmission and switching are used to balance the approximate number of ports and in the end to determine the best area. An rigorously exact value for the gain is not needed, because the evaluation is only used to compare relative gain among the possible areas.

4.4.2 Routing

When both network topologies –electrical and optical layers- are known, it remains to route the connections in the two layers. The routing can be divided into, three parts: starting SDH/SONET area, WDM backbone, ending SDH/SONET area. Those routings are quite independent, allowing some optimization in each part but the planner has to be careful not to break the strong correlation created between connections in the area-to-area part.

4.4.3 TDM Grooming

The composition of wavelength connections depends on the SDH/SONET switching capabilities. It means that a major part of the TDM grooming is done in the SDH/SONET domain . Hence the efficiency of TDM grooming in the WDM part is limited by the capacity of demultiplexing and re-arranging granularities with enough fineness at the SDH/SONET-WDM gateway. For instance if the gateway can not manage SDH/SONET connections at granularities smaller than STM1/STS3, those connections have to be composed with smaller connections going to the same area-destination. Otherwise multihopping will be needed to re-arrange the small connections implying more resource.

The TDM grooming process looks at each sub-path in the topology and aggregate SDH/SONET connections successively on the best sub-path in terms of number of ports saved. The same operation is done until all connections are groomed. The wavelength traffic matrix obtained is composed with a minimum of λ-connections staying in the optical layer as long as possible. All SDH/SONET connections are carried by a wavelength, but non-indispensable multihopping is avoided. Hence the global number of ports (electrical + optical) is reduced. The fig. 9 represents a wavelength connection composed with SDH/SONET connections starting or demultiplexed at the first node, going through several other nodes without multihopping and ending or demultiplexed/remultiplexed at the last node.

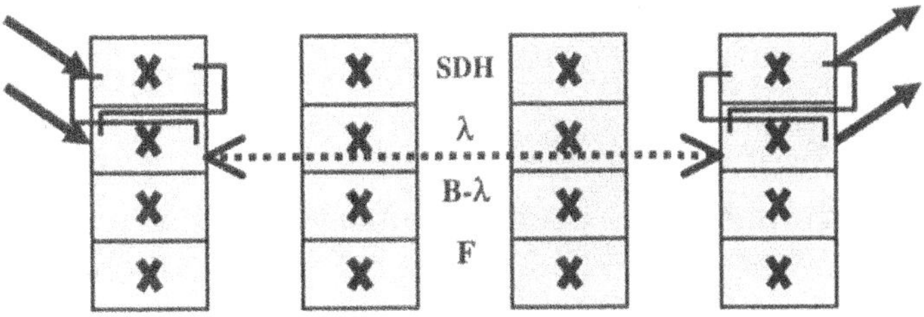

Figure 9. TDM grooming

4.4.4 WDM Grooming

The WDM grooming is done using the same principle as the TDM grooming. The process tries to avoid as much as possible the "optical" multihopping, that is the demultiplexing of band in wavelengths or fiber in bands.

First the process allows to multiplex the wavelength into bands. For each sub-path the number of ports saved by multiplexing wavelength in bands is calculated. The best is selected with the concerned connections and the operation is repeated until all wavelengths belong to a band.

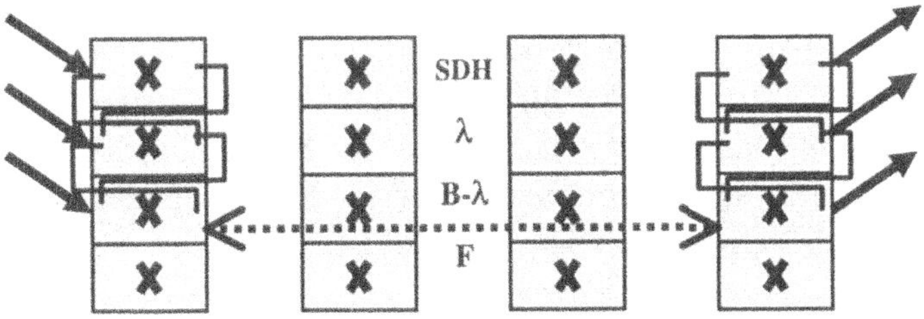

Figure 10. Grooming wavelength in bands

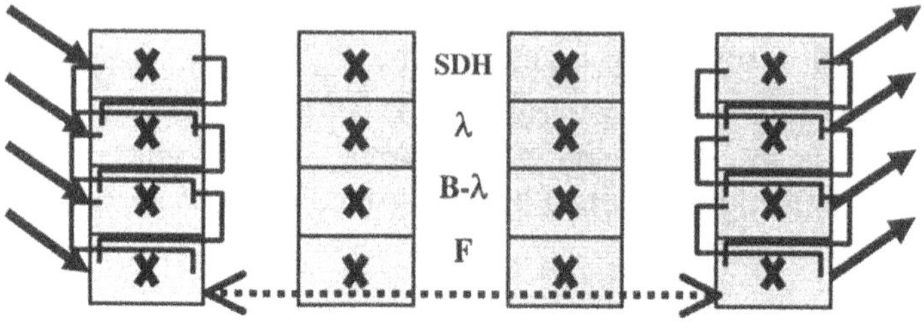

Figure 11. Grooming bands in fibers

Finally, Bands of wavelengths are groomed in fibers using a similar approach. Figures 10 & 11 represent the possible composition of a Band or Fiber connection and the transit along the path through others nodes without multihopping.

At each level, when multiplexing in a higher granularity, it remains some connections that do not fill completely the higher granularity. Below a given filling ratio, those connections can be re-routed in order to optimize the resource utilization ratio.

Finally, the outputs include the planning for each level. It gives the number of ports for each sub-node taking into account ports due to direct transit, ports due to internal add/drop between sub-nodes and ports due to direct add/drop.

Protection was not discussed so far. The introduction of several layer in the optical domain may justify some new schemes for protection. For instance like any connection, a band of wavelengths can be protected. In this case, while the traffic is already protected, it is no longer necessary to apply a protection scheme for each wavelength. This can be chosen for an objective of simplification for the global protection scheme. Anyway the protection scheme applied for a given level has to be done before the multiplexing into the higher level. Hence the resources dedicated to protection are also groomed and participate to the reduction of cost.

4.4.5 Final Resource assignment

Almost all the planning is done after the previous steps. It remains to allocate the resource with the classical approaches. In consequence of the introduction of wavelengths-bands, some band translations may be necessary in the network. This problem is entirely analogue to the classical wavelength conversion issue.

The gain expected from the introduction of several granularity in the optical layer can be improved with some dedicated algorithms integrated in the overall planning process. Some further analysis are expected to show that more available granularities in the optical layer can participate to more flexibility and scalability of the network. But the first argument to the introduction of this technology is without doubt, the cost gain obtained by the reduction of the number of ports in the network.

5. CONCLUSION

This introduction to Multi-Granularity Optical Networks discussed the technical availability, the opportunity and finally a methodology relating to one of the most advantageous features of optical technologies. The aptitude to switch various granularities in a flexible and cost-effective way is effectively a promising perspective to build ultra-high capacity optical networks.

In the first part a realistic MG-OXC architecture has been proposed with possible associated technologies. Then a first evaluation of switching requirements versus the increase of traffic showed that high-granularities are a possible solution facing the traffic evolution in a typical meshed network. Finally a dedicated planning and design process was described in order to optimize the benefits gained from the introduction of the concept.

This global approach gives the ability to go beyond simple incremental wavelength-based optical network and designs efficiently optical networks facing the traffic explosion without cost explosion.

6. REFERENCES

[1] "1.5 Tbit/s WDM transmission of 150 channels at 10 Gbit/s over 4x100 km of TeraLight fibre", S.Bigo et al., ECOC'99, PD2-9.

[2] "1 Tbit/s (100x10.7 Gbit/s) transoceanic transmission using 30 nm wide broadband optical repeaters with Aeff-enlarged positive dispersion fibre and slope compensation DCF", T.Tsuritani et al., ECOC'99, PD2-8.

[3] "1.6 Tbit/s (40x40G Gbit/s) transmission over 4x100 km non-zero-dispersion fibre using hybrid Raman/erbium-doped amplifiers", T.N. Nielsen et al., ECOC'99, PD2-2.

[4] "performance of a 576x576 optical cross-connect", H.Laor et al., NFOEC-99, 4th paper of B2 session.

[5] "Free-space micromachined optical cross-connects: routes to enhanced port-count and reduced loss", L.Y.Lin et al., OFC'99, paper FH1

[6] "A compact, scalable cross-connect switch using total internal reflection due to thermally-generated bubbles", J.E.Fouquet vet al., LEOS 98, paper ThS1

[7] "Dense Wavelength Division Multiplexer and Router using Diffraction gratings", J.P. Laude and al., NFOEC'99, Chicago,pp83-87.

[8] "Integrated all-active Mach-Zehnder wavelength converter with record signal sensitivity and large dynamic range at 10 Gbit/s", C.Janz et. al., OFC'99, paper FB3.

[9] "Ultra High Capacity Optical Transmission Network", Final Report of Action COST 239, 99.

[10] "Current European WDM Deployment Trends", E. Lowe, IEEE Communication Magazine, Feb. 98.

[11] "WDM: North American Deployment Trends", J. P. Ryan, IEEE Communication Magazine, Feb. 98.

[12] "Key Building Blocks for High-Capacity WDM Photonic Transport Networks", A. Jourdan and al., IEEE Journal on Selected Areas in Communications, vol. 16, no 7, Sept. 98.

[13] "Broadband Network Infrastructure of the Future: Roles of Network Design Tools in Technology Deployment Strategies" B. Doshi, P.Harshavardhna, IEEE Communication Magazine, May 98.

Network Architectures in Synergy with Ultra-wideband WDM Multiplexing Solutions towards a 1,000 Wavelength Channel Network

Alexandros Stavdas and Yannis Angelopoulos
Department of Electrical and Computer Engineering, National Technical University of Athens, Zografou 157 73,Athens, Greece,
tel+30-1-77 22 057, fax +30-1-77 22 076, e-mail: astavdas@cc.ece.ntua.gr

Key words: wide-band WDM, switching, routing, WDM multiplexer, WDM demultiplexer, access network, transport network.

Abstract: The paper presents the design issues for the development of networks exploiting recent advances and trends in optical technology allowing large numbers of WDM channels. The targeted topology consists of non-hierarchical all-optical rings made up by WDM PON networks and suitable optical cross-connects. The resulting benefits in terms of node consolidation, and increased capacity are analysed for a MAN/WAN network employing transmission and routing in the spectral window 1450-1650 nm and covering both residential and business customers. A single multiplexer/demultiplexer for the entire spectral range is presented and its system performance is calculated.

1. INTRODUCTION

The current deployment of WDM systems mainly targets capacity upgrades of point-to-point links using optoelectronic regeneration on both ends. As part of a gradual network evolution, the concept of all-optical networking proliferates by force of its ability to reduce the processing associated with transit traffic and the possibility to perform network functions like protection at the optical layer. At the same time novel requirements for a less hierarchical network composed by fewer network elements are emerging in response to the never ceasing quest for reduced deployment cost and increased network reliability. In particular, the ongoing explosion of Internet related services have stimulated the generation of network concepts where packets are directly interfaced to the optical layer. In this context, the most critical factor is the number of available wavelength channels that allows to simultaneously meet the demand for both capacity enhancements and flatter network structures. Non-hierarchical networks based on a large number of wavelength channels are a subject of intense research effort [1-2]

Regarding the issue of capacity upgrade, currently there are two main approaches. In the first approach the systems are forced to operate at an increasingly finer channel spacing within the gain bandwidth of the EDFA. In the second approach, the number of wavelength channels is traded for a higher bit-rate per channel. In both cases the product *(number of channels) x (bit-rate/channel)* remains constant limited by the current amplifier gain bandwidth. Particularly in the former case the associated requirements in terms of wavelength stability, crosstalk, physical dimensions of the multiplexers/demultiplexers, non-linear effects in the fibre etc. are quite stringent. The inefficacy is due to the striking mismatch between the low attenuation spectral range centered around the 1550nm wavelength which is approximately 200nm, and the gain spectrum of the EDFA which is limited to a maximum of 35nm. Recently efforts had been made to utilise both the C and L bands of the Er^{+} doped amplifiers [3]. Simultaneously, the renewed interest for Raman amplifiers as well as the prospect of using SOAs as in-line amplifiers [4] has opened-up the possibility to envisage transmission (and routing) beyond the C and L bands of the Erbium optical amplifiers. Under any circumstances, one of the next major developments of optical networks will encompass all-optical transmission and routing in the spectral range 1450-1650 nm towards the ultimate goal of 1280-1680 nm.

So far, studies have been presented which tackle parts of such technology considerations [5-8]. The aim of this paper is to present the calculated performance of a wavelength multiplexer/demultiplexer (mux/demux) in the range 1450-1650 nm and to highlight the essentials of a network architecture that is based on such a large number of WDM channels.

In section II, a generic partitioned network structure based on non-hierarchical rings that could radically modify the way trunk and access networks are perceived is illustrated. Also, a possible OXC architecture that provides an interface between the two network classes i.e. access and trunk is presented. In section III, the derivation of the performance of the mux/demux is presented and its system implications are identified.

At this point estimation on the total number of wavelength channels has to be given. Wavelength multiplexers/demultiplexers with channel spacing 0.2-0.5 nm have been reported [5, 9]. Assuming that this constant channel spacing is feasible across the entire spectrum (section III), the number of wavelength channels could range between 400 and 1,000.

2. NETWORK ISSUES

2.1 Network structure

Network designers can capitalise on a vast increase in the number of available wavelength channels to generate a less-hierarchical network layout with significant node consolidation and reduced processing of transit traffic. The cost of transmission compared to the cost of extensive tandem switching is continuously dropping indicating that node consolidation could be the means of scaling down the cost of providing broadband services to the customer. At the same time levels of the traffic aggregation hierarchy, particularly in MAN networks can be eliminated. In figure 1 such a partitioned network is illustrated that has as a basic building block a ring network. A ring network is a well-known network topology with excellent protection/restoration properties and is the predominant choice in most deployed networks. The physical extension of each ring incorporates a cluster of nodes. Clustering helps to localize any network fault, and it scales down wavelength channel requirements (via wavelength re-use in physically different rings).

Of particular interest is the bidirectional dual ring (i.e. with a clockwise and counter-clockwise traffic flow). The ring networks shown in figure 1, could be either single or multiple dual fibre. The interconnection between the rings is achieved via suitably designed optical crossconnects (OXC) located on a shared node between two (or more) rings. Apart from this common node, additional physical connectivity is provided via direct links resulting in a partially meshed network that reduces the number of hops that needed to reach the final destination. Providing interconnection between MAN/WAN rings allows constructing larger networks. In the following section the rationale in integrating a MAN with a PON-based access networks is presented.

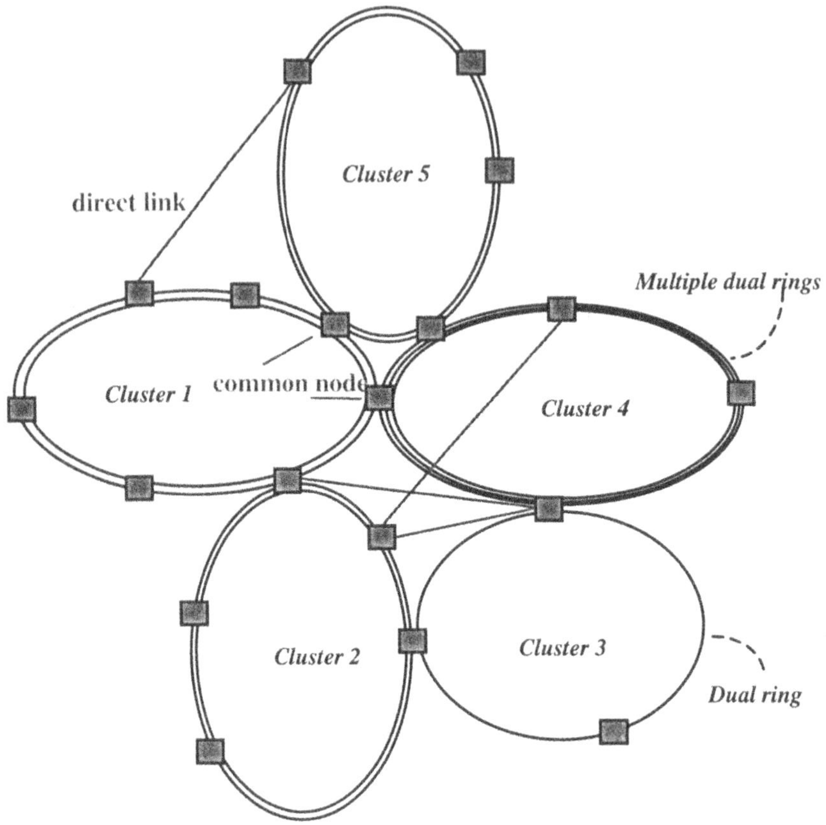

Figure 1. [A partitioned network based on rings]

2.2 Dimensioning of an ultra-wideband WDM PON

The overall architecture of a single ring is presented in more detail in figure 2. There are 4 nodes interconnected via multiple dual rings. Each node is attached to a WDM PON like the one described below that form the access network. The traffic generated within each PON can be connected to the other nodes either transparently via direct access to the OXC or via an IP router, ATM switch or SDH crossconnect.

The ultra-wideband WDM PON is formed by means of a WDM demultiplexer (downstream traffic) or multiplexer (upstream traffic) that covers the spectral range 1450-1650 nm. The proposed architecture exhibits fundamental differences compared to the established PON structure.

In a traditional PON which is based on power splitting, a single wavelength is divided in successive splitting stages in such a way that all ONUs receive copies of the same signal. The power splitting PON will be the first to be installed avoiding the cost of installing a transceiver pair for each customer, since the OLT is shared by all. The sharing of the bandwidth among hundreds of customers in a power splitting PON takes place by means of TDMA multi-access protocol techniques. This at the same time carries out traffic concentration, which is a further source of cost savings. The power splitters create a broadcast medium in the downstream requiring encryption for the support of point-to-point services.

Very different is the traffic situation in the upstream direction. The transmissions from all ONUs now converge to the shared feeder fiber creating the need for access arbitration. Each slot contains apart from the payload a physical layer preamble for the burst mode operation. Since each transmission comes from a different transmitter and a different distance, several physical layer problems must be overcome. To make slots unambiguous to all units, the ONUs are directed to add an electronic delay that would bring them all at the same virtual distance. The amount of delay, which is inversely proportional to the ONU distance, is identified by means of a ranging process and then communicated to the ONU. Once a slotted transport medium has been created, it is the role of the Medium Access Control (MAC) to arbitrate the access to the slots. The slot allocation is based on access permits sent downstream, which identify the ONU taking the next slot. The MAC protocol is typically of a dynamic reservation type and not semi-static [10] to respond to the temporal fluctuations of traffic that can be quite dramatic in the broadband services.

In contrast to power splitting, the wavelength splitting can offer much larger bandwidth to the final customer. In this system, a particular wavelength channel is always directed towards a prespecified PON branch passively but without power splitting. This can be achieved using an *1:M* wavelength demultiplexer, *M* being the total number of wavelength channels. The demultiplexer connects an input wavelength to a particular output fibre. Assuming a minimum of 400 wavelength channels, power budget degradation can be avoided since whilst a 400-fold passive splitting would have resulted in 26 dB losses the demultiplexer losses can be stay below 5-7 dB maximum (section III). Therefore, a larger number of customers could be serviced assuming the same power budget. For the necessary power equalisation, band-optimised optical amplifiers can be used like in [3]

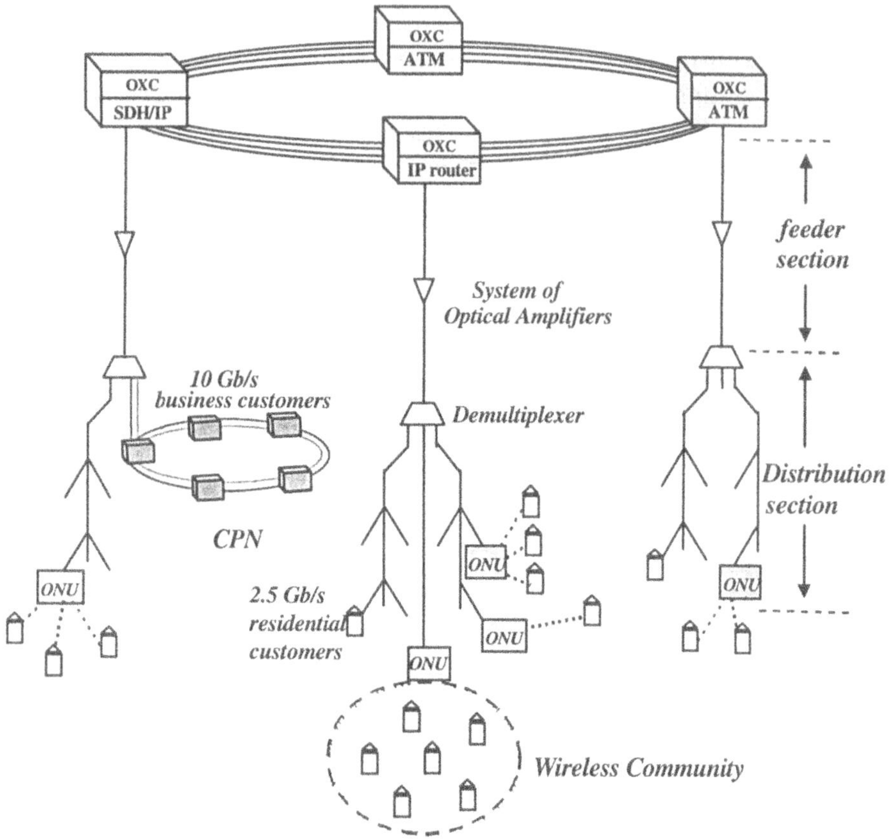

Figure 2. [A generic architecture for the proposed ring]

In the downstream direction each wavelength, after the demultiplexer, is related to one single-wavelength PON. However, the power splitting part is limited and the same wavelength is terminated in a small number of optical network units (ONUs). A number of customers could be connected to a single ONU using various techniques, like xDSL, which leaves in a first stage the copper based home equipment intact delaying the added costs of optics in the home. Customers with higher bandwidth needs e.g. teleworkers or professionals or even small offices can be placed in a channel shared by fewer TDMA users and can be upgraded to fibre-to-the-home (FTTH). Even medium businesses can eventually be upgraded to a separate WDM channel instead of TDMA sharing. The additional advantage of the separate wavelength apart from the abundant bandwidth is also enhanced security and connectivity since a particular wavelength would be passively routed to a specific destination and the contained information would not be broadcast to all customers. These wavelength channels could operate at 2.5

Gb/s. In the upstream direction further modifications to MAC related hardware has to be introduced to achieve these high speeds. However, the necessary technology is already developed e.g. Gigabit Ethernet.

The solution shown in figure 2 illustrates an advanced infrastructure that addresses simultaneously both market segments i.e. business and residential customers. The total number of wavelength channels (>400) is divided into two groups. The first subgroup of channels is destined to the *residential* customers. The other subgroup serves the high capacity *business* users (CPNs, ISPs, wireless LANs etc). The *residential* users are connected in the network in a non-transparent mode whilst transparent connections serve *business* customers. Different scenarios could be envisaged for channel splitting between residential and business customers depending on traffic patterns, population density, demand for new services etc. Here it is presumed that from the 400 wavelength channels per PON, 300 channels are allocated to residential customers and 100 channels are allocated to the high capacity users.

Assuming that on average 2 Mb/s fully symmetrical bandwidth is offered for interactive services to each residential customer, then 1,250 customers can be served by this single-wavelength power-splitting PON. The overall access system could be seen as formed by 300 single-channel power-splitting PONs, one per wavelength. Consequently, it can be concluded that 375,000 residential customers can be served from the overall access system. In this case, a total capacity of 750 Gb/s is allocated to residential customers imposing the need for node consolidation and implying that Terabit routers or crossconnects with a large number of ports (>300) should be accommodated to the central office as it is shown in fig.2.

The business customer channels are not crossconnected via the SDH/ATM switch. These channels are connected in a transparent mode to the trunk network via the OXC. This means they have direct access to the OXC and, therefore, a "clear channel" is provided. The bit rate per channel for the business customers can be at 10Gb/s or even higher.

2.3 Dimensioning of the MAN network

The nominal transmission rate of each channel in the trunk network depends on many parameters. Namely, service demand and number of wavelength channels available as well as on the trade-off in cost between adding a new wavelength channel or increasing the bit rate in order to achieve a given throughput. In any case, transmission at 10-40 Gb/s, or even at higher rates could be assumed. For the residential customers, it is expected that grooming and bandwidth aggregation can be provided by the Terabit IP router/SDH DCC. Hence for residential customers, 75 wavelength at 10 Gb/s should be allocated in each node.

These wavelengths are destined either to nodes within the ring or to nodes located in other rings. The multiple fibre ring gives the option to direct the particular traffic to the final destination either using transparent connections or via the adjacent "digital electronic switch" through o/e conversion in a multi-hop scheme. All these wavelength channels are routed within the network via the OXC. These channels must have a different nominal wavelength from the one used by *business* customers. Obviously, any wavelength channel can be present only once in a given link. Therefore the traffic emanating from a particular node should be transported using either a different subset of channels or via another fibre. In the former case, the total number of nodes in a dual fibre ring depends on the total number of wavelength channels available. Assuming, 400-1000 channels it follows that 4-10 nodes can be placed per dual fibre.

Now, one can estimate the total number of dual fibres needed per ring. A reasonable minimum is one dual fibre for residential and one for business customers for connections within the ring, one dual fibre for residential and one for business customers for connections outside the ring. Thus, a total of 4 dual fibres would be needed. In any case, these are indicative values. In addition one may envisage additional fibres for heavy transit traffic; the number of dual fibres would range from 4 to 20 per ring.

2.4 The proposed OXC architecture

From the aforementioned analysis, it is emerging that an OXC architecture capable to crossconnect a large number of wavelength channels between a relatively small number of fibres is sought. In the proposed ring structure, the OXC is co-located with the DXC (SDH/ATM/IP router). Here we briefly present an architecture that is particularly suitable when there is a large number of wavelength channels per link [11].

It is named "wavelength crossconnect" (WXC) and it is illustrated in fig.3. N input/output fibres are assumed with M wavelength channels per fibre. The multiwavelength signal from each incoming fibre is amplified before power split N times at the input stage and directed towards a wavelength selector. The wavelength selector can select an arbitrary number of wavelengths ranging from 0 to M and it is the main building block of the crossconnect. Entering the selector, the wavelength channels are demultiplexed and each wavelength is either forwarded (on-state) or blocked (off-state) towards the multiplexer from a dedicated device. This device can be a SOA gate, a modulator or any other device that is capable of generating a large extinction ratio. The most promising candidate is a SOA gate since it has a significant on/off ratio, it can provide gain, and it also provides fast reconfiguration [12]. The selected channels after multiplexing are directed towards a *N:1* power combiner attached to an output fibre. A wavelength channel can be switched from any input fibre to any output fibre when the appropriate gate in the dedicated selector is activated. An important characteristic of the architecture is that only one out of *N* gates per wavelength channel destined to the same output fibre is activated at any time.

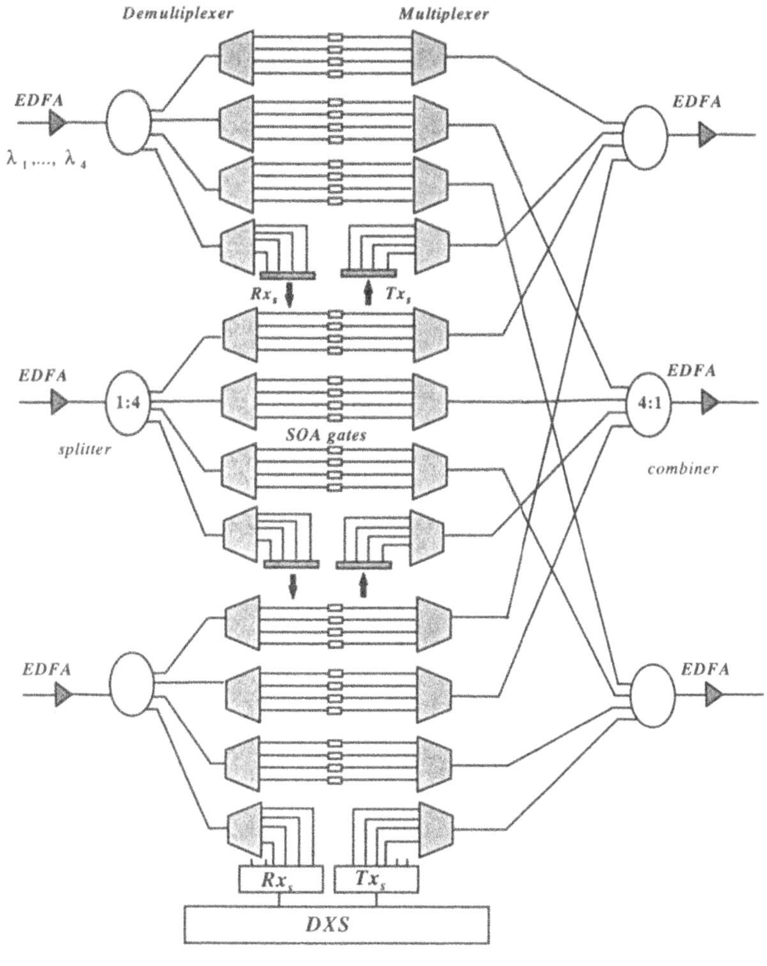

Figure 3. [The OXC architecture.]

The proposed architecture integrates space and wavelength switching. There are no crosspoints or crossovers between any two wavelength channels since wavelength switching from a given input fibre to the requested output fibre is achieved with the activation of a single gating element (e.g. a SOA gate). The performance of the proposed architecture is independent on the number of the wavelength channels and the architecture is modular with respect to wavelength and link. The total number of gates in MN^2, the components scale linearly with the number of wavelengths.

There are two sources of crosstalk in the WXC architecture. The hetero-wavelength crosstalk due to imperfect demultiplexing between adjacent channels, and the homo-wavelength crosstalk due to imperfect extinction of the SOA gates that operate at the same nominal wavelength from the other *N-1* input fibres. Also, the hetero-wavelength crosstalk due to imperfect demultiplexing can be a source of homo-wavelength crosstalk at the output fibre. Fortunately, since the diffraction gratings in the wavelength selector operate in tandem, the overall adjacent channel rejection is twice the rejection of a single multiplexer/demultiplexer. Thus, this crosstalk either seen as hetero-wavelength or homo-wavelength is significantly suppressed. Homo-wavelength crosstalk is well mastered due to the high on/off ratio of a SOA.

Here the node cascadeability is assessed against the power penalty due to homowavelength crosstalk. In the current case, a small number of links (4-20) is envisaged. Two extreme cases are examined. A node with 4 input fibres (i.e. 3 sources of homo-wavelength crosstalk at the output) and a node with 20 fibres (19 homowavelength interferes). In both cases the power penalty is calculated by the statistical method used in [13]. The power penalty for an AC coupled receiver is defined as

$$\Re = -10\log_{10}\left(1-6\cdot\sqrt{\sum_{i}^{N}\varepsilon_i}\right) \qquad (1)$$

where ε is the ratio $P_{sig}/P_{crosstalk}$ for a single interfere. It is assumed that all crosstalk sources from all nodes have the same coefficient (ε) which is -50 dB [12]. In figure 4 the power penalty is calculated as a function of nodes to be cascaded assuming 4 input/output fibres per node (solid line) and 20 fibres per node (dashed line).

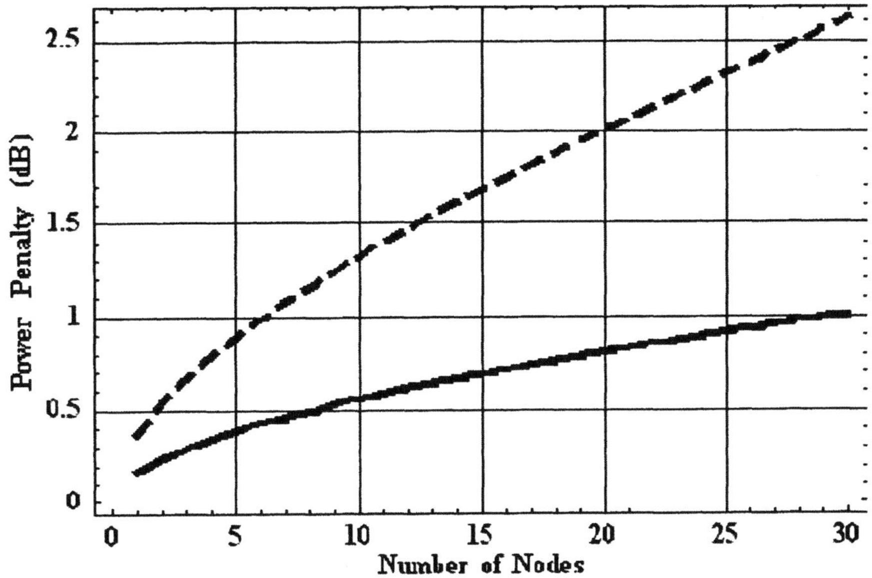

Figure 4. [Power penalty v. number of nodes for 4x4 I/O fibres (solid line) and for 20x20 I/O fibres (dashed line)

One is tempted to calculate the throughput of the WXC. For a 4 I/O fibre WXC, 400 channels at 10 Gb/s per channel, the capacity is 16 Tb/s whilst for 20 I/O fibres this capacity is 80 Tb/s. In the proposed architecture to total number of ports is analogous to the number of wavelengths. For a large number of wavelength channels (M>400), the number of ports escalates to prohibitive numbers. One way of dealing with this problem is to consider the concept of wavelength bundle routing [8]. At the same time upgrading the bit rate of a wavelength channel for example from 2.5 Gb/s to 40 Gb/s can more easily be accommodated when a hierarchical switching concept is adopted.

In the proposed architecture if the wavelength bundle is consists of wavelengths that are spectrally displayed by one grating order (e.g. $m\ \lambda_1$, $(m+1)\ \lambda_1$, ..., m is the grating order), then the entire bundle can be switched by the same gating device. In this way, the total number of ports will be reduced by a factor proportional to the number of wavelengths in the bundle. Partitioning the I/O fibres of the OXC for servicing exclusively transit and local traffic allows for a significant reduction in the total port count.

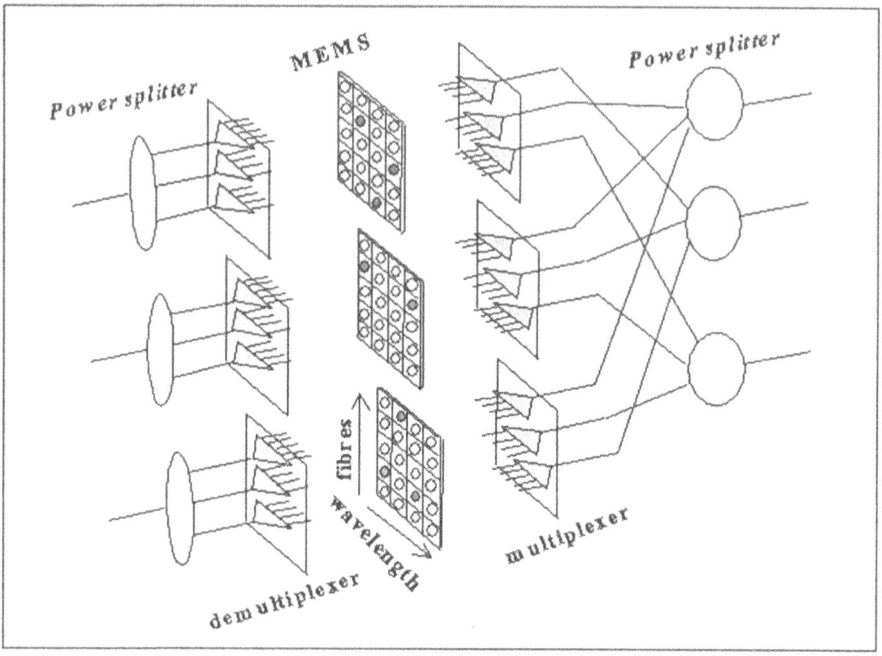

Figure 5. [The WXC architecture implemented with free-space concave holographic grating and a 2-D array of MEMS

Further modifications to the WXC architecture, to tailor advancements of optical technology, are feasible in order to reduce cost and simplify the overall operation. One advantage of free space concave gratings is that allows taking advantage of the massive parallelism offered by free-space optics. It has been shown experimentally [14] that more than 20 fibres can be placed on a plane vertical to the plane of dispersion using free-space concave gratings to acquire up to 20 demultiplexers in parallel. Hence, for relatively low number of I/O fibres, a single device can replace the *N+1* demultiplexers illustrated in figure 3. Using the MEMS technology [15], a 2-D array of 20x400 (or more) of "shutters" can be, eventually, implemented. The modified WXC architecture is illustrated in figure 5.

3. OPTICAL TECHNOLOGY ISSUES

A free-space concave grating has been identified as a key device in spectrally extended wavelength routed networks [5, 14]. As it is pointed out in [5, 16], the aberration corrected spectrum of these devices exceeds 100 nm and this additional bandwidth can be used for wavelength routed purposes.

In contrast, integrated optic devices like AWGs or two dimensional concave gratings, have inherent limitations to cover this extended wavelength range since they operate at high diffraction orders which in turn limits the Free-Spectral-Range [9]. Despite that, WDM transmission of 200nm based on AWGs can be achieved using a cascade of band selective filters. Effectively, this approach will lead to a sort of hierarchical multiplexing/demultiplexing something that will complicate wavelength routing concepts, especially the construction of a viable OXC, it will increase the overall cost, and it will degrade the power budget of the link.

Here the coupling efficiency and crosstalk performance of a single free-space holographic concave grating listed in [5], Table I is presented. This concave grating has 10 nm/mm linear dispersion. It is pointed out that this demultiplexer has been optimised for best performance at 1550 nm.

Table 1. [Parameters of the holographic concave grating.]

Radius of Curvature R	347.14368 (mm)
Distance of source r_A	344.30981 (mm)
Distance of main image r_B	350.00977 (mm)
Angle of incidence α	18.011561 (degrees)
Angle of diffraction β	7.4629230 (degrees)
Recording angle γ	44.195223 (degrees)
Recording angle δ	33.977113 (degrees)
Recording wavelength λ_o	488 (nm)
Distance of recording source C	926.5228 (mm)
Distance of recording source D	740.3296 (mm)

In Table I, accuracy up to seventh digit has been maintained to highlight the sensitivity of the computational method. Using these values, system related parameters like coupling efficiency and heterowavelength crosstalk can be calculated. Both coupling efficiency and crosstalk are calculated using the overlapping integral between the input and output single mode fibres [17, 18] based on the analytic expression of the wave aberration function as described in [5].

The coupling losses due to diffraction effects are calculated to be 0.8 dB. One should add another 1.5 dB of losses due to the diffraction efficiency of the grating (for a holographic grating, efficiency curves are typically 70%-75% for both polarisations in the spectral range of interest). In figure 6 the coupling losses are calculated across the wavelength range 1530-1570 nm. It can be observed that the losses are practically constant across the whole spectrum. The overall losses include losses due to residual aberrations, estimated grating efficiency and diffraction effects due to the grating aperture.

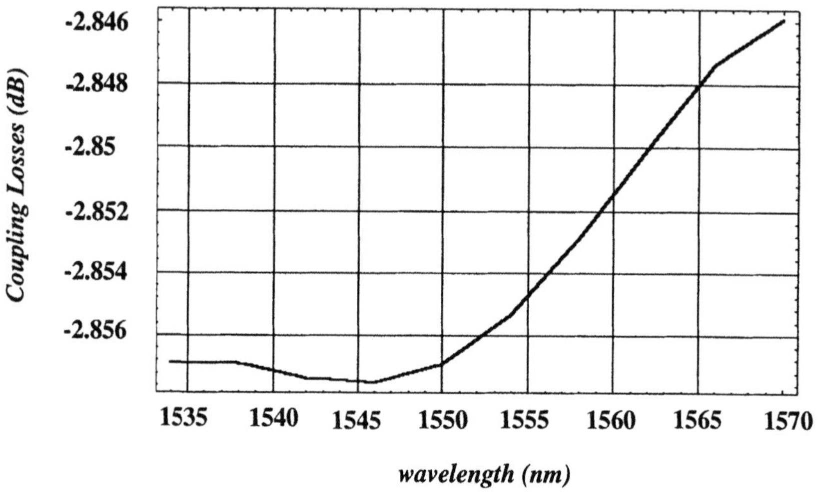

Figure 6. [The coupling losses of the holographic grating in the band 1530-1540 nm.]

The coupling losses across the wavelength range 1450-1650 nm are shown in figure 7.

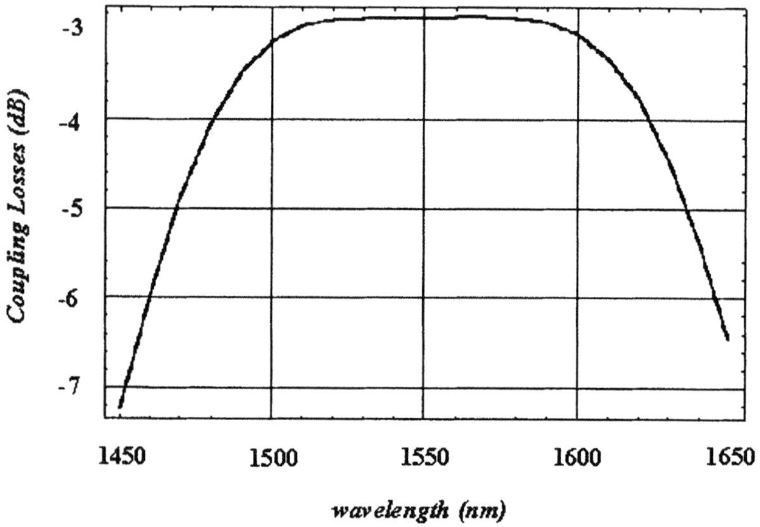

Figure 7. [The coupling losses of the holographic grating in the band 1450-1650 nm.]

Apparently immediate improvement of the coupling loss figure of 3 dB is achievable with a minor modification of the holographic recording scheme. Further design improvements are possible. The crosstalk at the immediate neighbors of λ=1550nm is illustrated in the figure 8. The crosstalk at the first neighbor is -53.1 dB.

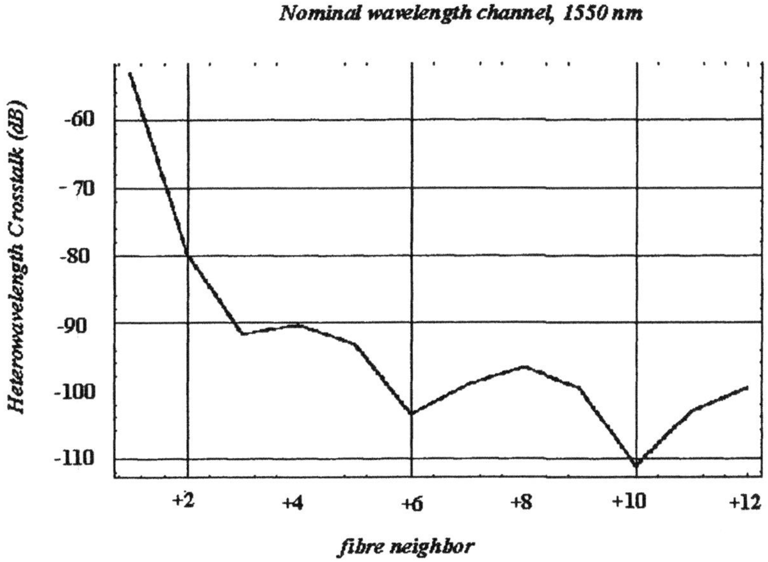

Figure 8. [The crosstalk at the neighbour channels due to the channel at 1550 nm.]

The crosstalk at first immediate neighbours due to λ=1570nm is illustrated in the figure 9. The exact value for the first neighbour is -52.1 dB.

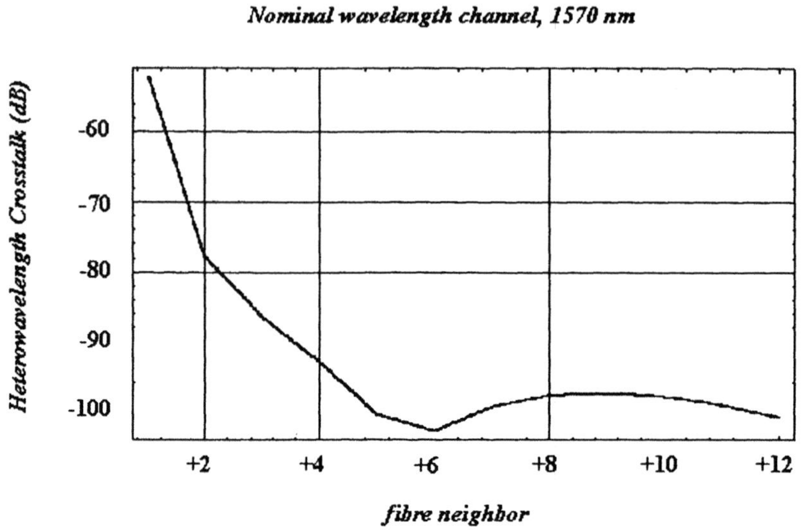

Figure 9. [The crosstalk at the neighbor channels due to the channel at 1570 nm.]

Remarkably, the crosstalk at the first neighbor due to λ=1450nm is -47.9 dB. The crosstalk at the first neighbors is illustrated in figure 10. A similar curve is obtained for λ = 1650 nm.

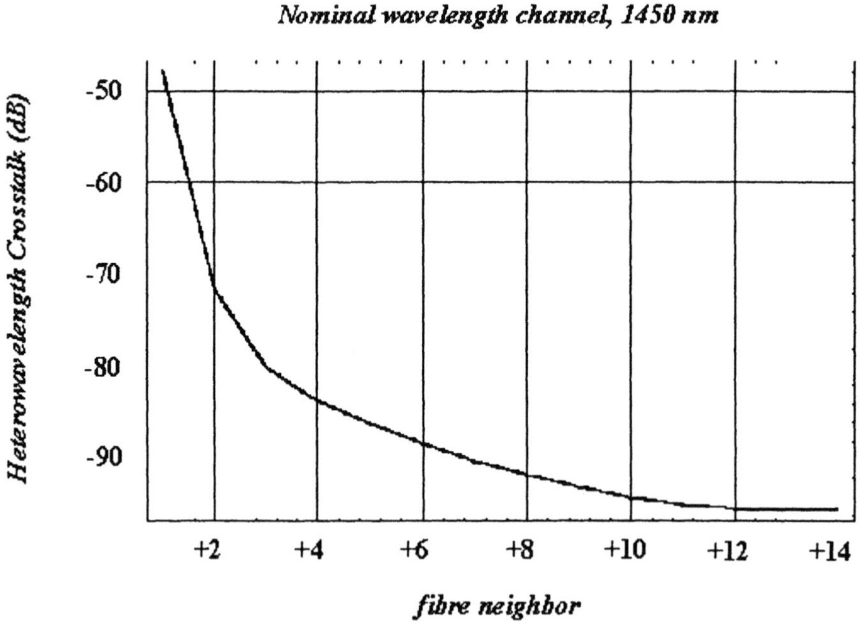

Figure 10. [The crosstalk at the neighbor channels due to the channel at 1450 nm.]

It is mentioned again that this grating was optimised having in mind the EDFA's gain spectrum. From the above analysis, it is concluded that the same concave grating is capable of multiplexing/demultiplexing wavelength channels that spectrally cover 200nm of bandwidth. The total capacity of this grating is of 400 wavelength channels.

4. CONCLUSIONS

The implications in network structures when considering WDM transmission in the band 1450-1650 nm are enormous. Extraordinary capacity enhancements, node consolidation, elimination of many multiplexing layers and broadband accessing for large sections of the population are simultaneously feasible. We have presented some considerations regarding network architectures and we have studied the feasibility of these concepts in terms multiplexing technology and connectivity offered by wavelength crossconnects (WXC).

ACKNOWLEDGEMENTS

This work has been supported by the Greek Government through the GSRT 97EL-116 project and by ACTS Project AC355 PELICAN

REFERENCES

[1] P. Caponio, A. M. Hill, R. Sabella, "Switchless Optical Network for Advanced Transport Architecture", IEEE/LEOS '98, Orlando, FL, USA, December 1998

[2] R. Sabella, S. Binetti, M. Listanti, A. Maga, "Interconnection of "switchless" Optical Networks for the realization of a Large Scale Networks", ECOC '99, p. 428, Nice, France, September 1999

[3] Y. Sun et al. "80 nm ultra-wideband erbium doped silica fibre amplifier", Electronics Letters, Vol.33, p.1965-1967, 1997

[4] L. H. Spiekman et al, ECOC99, PD2-7 p.36, Nice, France, September 1999

[5] A. Stavdas, P. Bayvel and J. E. Midwinter: Design and performance of concave holographic gratings for applications as multiplexers/demultiplexers for wavelength routed optical networks, Optical Engineering, vol. 35, 1996, pp. 2816-2823.

[6] M. Jinno et al: "Design of ultra-wide band (1450-1650 nm) WDM transmission systems considering intra- and inter wavelength band nonlinear interactions", NOC'98, Technology and Infrastructure, p.205-209, (1998)

[7] K. Lee et al:, "Multiwavelength all optical networks with wavelengths outside the Erbium doped fibre amplifier bandwidth", Journal of Lightwave Technology, Vol.13, p.791, 1995.

[8] M. Jinno, J. Kani, K. Oguchi: " Ultra-wide-band WDM networks and supporting technologies, NOC'99, Core Networks and Network Management, pp. 90-97, 1999.

[9] K. Okamoto K. Syuto, H. Takahashi, Y. Ohmori: Fabrication of 128-channel arrayed-waveguide grating multiplexer with 25 GHz channel spacing, Electronic Letters, vol. 32, no. 16, pp. 1474-1476, 1996

[10] J. D. Angelopoulos, N. I. Lepidas, E. K. Fragoulopoulos, I.S. Venieris,"TDMA multiplexing of ATM cells in a residential access SuperPON", *IEEE Journal on Selected Areas in Communications,* Special issue on high capacity optical transport networks, Vol. 16, No. 7, September, 1998. S

[11] A. Stavdas, H. Avramopoulos, E. Protonotarios, J. E. Midwinter: " A novel architecture for a wavelength crossconnect node", Photonic Network Communications, Vol. 1 (1), pp.77-88, 1999

[12] M. Renaud, M. Bachmann, M. Erman: Semiconductor optical space switches, IEEE Journal of Selected Topics in Quantum Electronics, vol. 2, no. 2, pp. 277-288, June 1996

[13] J. Zhou et al. "Crosstalk in multiwavelength optical crossconnect networks, Journal of Lightwave Technology, Vol.14, p.1423, (1996).

[14] F. Timofeev et al. "Free-space aberration corrected grating demultiplexer for application in densely-spaced, sub-nanometer wavelength routed networks", Electronic Letters. Vol.31, 1368, 1995.

[15] L. Yin, E. Goldstein, J. Simmons, R. Tkach, " High density micromechanical polygon optical crossconnects exploiting network connection-symmetry, Photonics Technology Letters, Vol.10, pp.1425-1437, 1998

[16] A. Stavdas: Design of multiplexer/demultiplexer for dense WDM wavelength routed optical networks, Ph.D. Thesis, University of London, 1995.

[17] R. Wagner et al. "Coupling efficiency of optics in single mode fiber components", Applied Optics, Vol.21, 2671, 1982.

[18] E. Churin et al. "The influence of aperture size and shape on crosstalk levels in free-space grating demultiplexer for WDM networks", Photonic Technology Letters, Vol.8, 1337, 1996.

Efficient Planning Tool for WDM Transport Networks

C. BUNGARZEANU, L. BESUCHET, D. RODELLAR
Ecole Polytechnique Fédérale de Lausanne
DE – Laboratoire de Télécommunications
CH-1015 Lausanne, Switzerland
E-maiL:Cristian.Bungarzeanu@epfl.ch

Key words: WDM, all optical networks, wavelength assignment, network planing

Abstract: Last generation optical components jointly with WDM techniques will allow the realisation of a switched optical layer based on wavelength routing of semi-permanent paths. This paper presents a computer-aided planning tool intended to cover the dimensioning, routing, and wavelength assignment in meshed networks with WDM optical cross-connects. An efficient graph-colouring algorithm allows a fast wavelength assignment even for topologies with a large number of nodes. The performances of the planning tool are illustrated on a 8-nodes European optical network.

1. INTRODUCTION

In response to increasing pressure on transmission capacity, brought about by upsurge in both narrow-band and emerging broadband services, many carriers are considering deploying dense wavelength-division multiplex (WDM) systems as a means of taking advantage of the fibre bandwidth [1]. In addition to point-to-point line systems, new optical elements with networking features, such as optical add/drop multiplexers (OADM) and optical cross-connects (OXC) have been tested in field trials.

These optical components of last generation allow for the manipulation of wavelengths in WDM networks offering advanced features such as wavelength routing, switching, and conversion/interchange [2]. In this way, protocol-transparent optical channels can be implemented between any source-destination nodes. Moreover, WDM techniques offer enough spare capacity for protection and restoration purposes. The design of WDM networks, including topics such as topology optimisation, routing and dimensioning, wavelength assignment, protection and restoration represents a complex optimisation problem.

In this context, there is a growing need for planning tools to deal with the main issues of optical networking in an automated manner. This paper describes a software application intended to address the planning of high-speed WDM networks. This software tool is currently under development so that several major planning issues (protection and restoration, transmission quality) will be dealt with in the near future.

2. NETWORK MODELLING

Nodes consist of optical cross-connects with the architecture depicted in Fig. 1. The number of output fibres M is the same as the number of input fibres. Each fibre carries L wavelengths. The optically transparent OXC is able to connect any wavelength from an incoming fibre to the corresponding wavelength of one of the outgoing fibre. It is assumed that there is no wavelength conversion within the OXC, so that there is no need of wavelength converters and the network management is simplified since only one wavelength is attached to an end-to-end channel. However, a blocking situation may occur if two channels having the same wavelength attempt to use the same outgoing fibre. Each OXC performs a static routing, switching occurring only in case a reconfiguration after a failure.

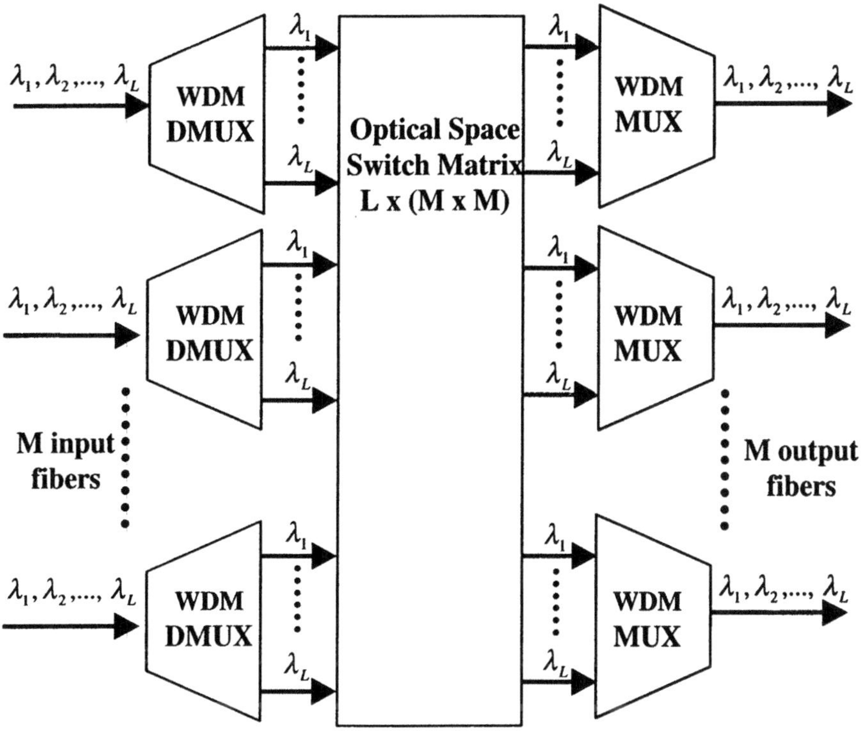

Fig. 1 Optical cross-connect

As shown in Fig. 2., a link *l* contains *k* fibres, each fibre supporting up to *L* wavelengths (*L* is constant over the network). Thus, the capacity of the link *l*, expressed in terms of optical channels is $C_l = kL$.

2.1 Routing and fibre assignment

Routing can be performed manually or automatically. A shortest path routing is available by running Floyd's algorithm [3], which computes simultaneously the routes for all connections. Two options are available:

- minimise the connection length
- minimise the number of intermediate nodes

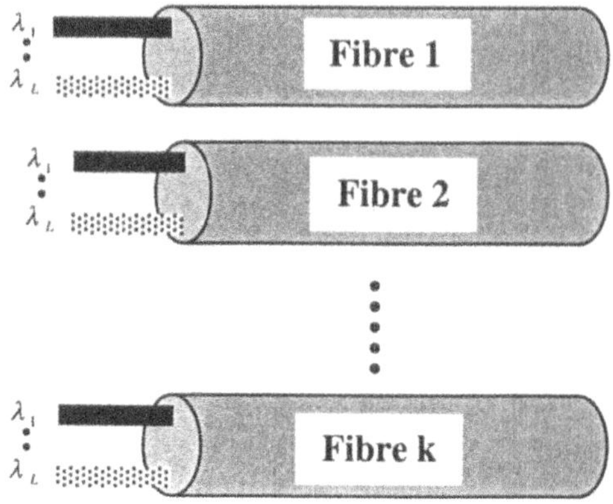

Fig. 2 Link structure

An automated fibre assignment is done for each link in the following way: the first channel is assigned to the first fibre, the second channel is assigned to the second fibre, the *k* channel is assigned to the *k* fibre, the *k*+1 channel is assigned to the first fibre and so on. In this way, the channels sharing the same link are equally distributed over the available fibres of the link. If the number of channels using a particular link exceeds the link capacity a warning message is issued to increase the number of fibres. At this stage, the wavelength assignment is not yet performed and even though the link capacity is greater than the number of channels that share it, it could be not sufficient to allow a global wavelength assignment solution. If, because of wavelength assignment restrictions, new fibres are added, the channels are redistributed to ensure an equal load of each fibre.

2.2 Wavelength assignment

The wavelength assignment is reduced to a graph colouring problem [4][5]. Each end-to-end channel is represented by a vertex in a neighbourhood graph. Vertices are connected by an edge if and only if the corresponding channels share some common fibre. The wavelength allocation can be mapped to a graph node colouring problem, which is a well-known NP complete problem. The heuristic method employed in our software is based on the Tabu search technique called TABUCOL [6]. Tabu search is a global search technique that moves step by step towards the minimum value of a function, which is in our case the number of neighbours (vertices connected by an edge) having the same colour. A tabu list of forbidden movements is updated during the iterations to avoid cycling and being trapped in local minima. In [6] it is shown that for graph colouring, TABUCOL is significantly superior to the simulated annealing.

The performance of the wavelength allocation technique was estimated on the network architecture shown in Fig. 3.

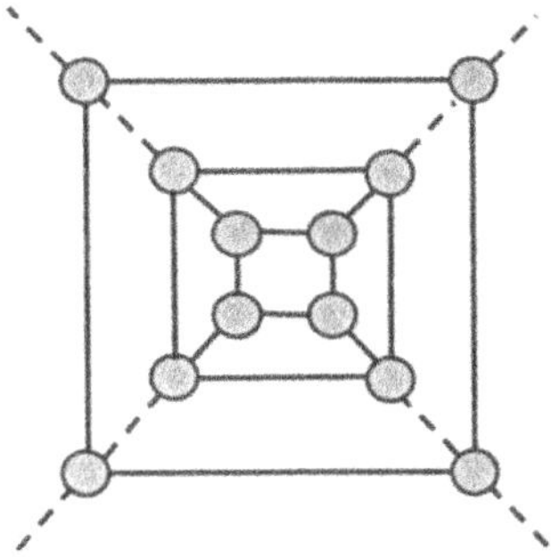

Fig. 3 Spider web test architecture

Assuming for simplicity only one fibre per link and one channel between each node pair, it turned out that the number of wavelengths per fibre L depends linearly on the number of node (see Fig. 4). However, as it is shown in Fig. 5, the number of iterations necessary to get a colouring solution grows exponentially.

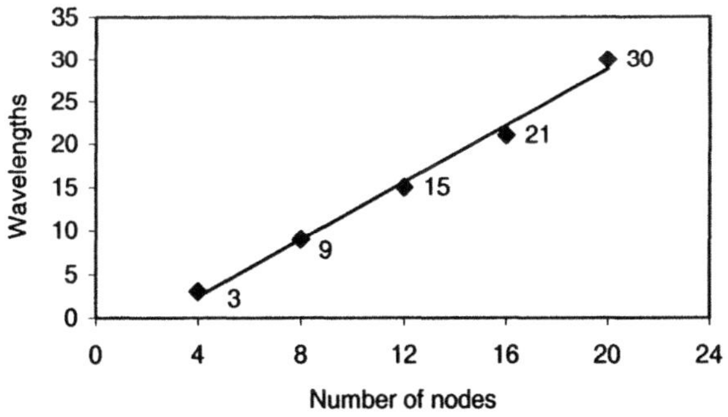

Fig. 4 Minimum number of wavelengths versus the number of nodes for the web network architecture

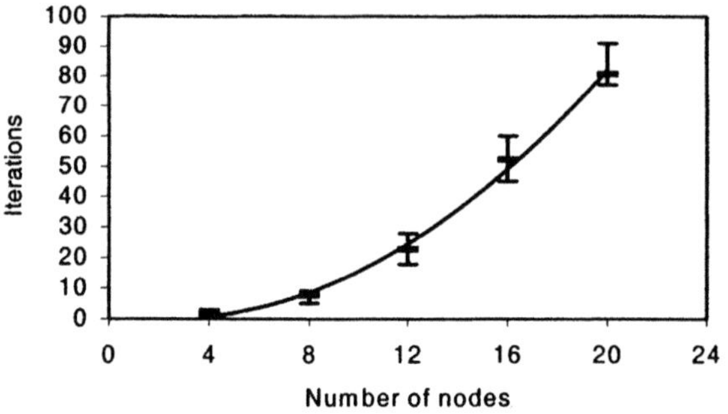

Fig. 5 Number of iterations necessary for a colouring solution

3. EUROPEAN TRANSPORT NETWORK

The planning tool described above has been used to design a 8–nodes central European network [7] the topology shown in Fig. 6. The distances between nodes and the demands expressed in terms of optical channels are listed in Table 1.

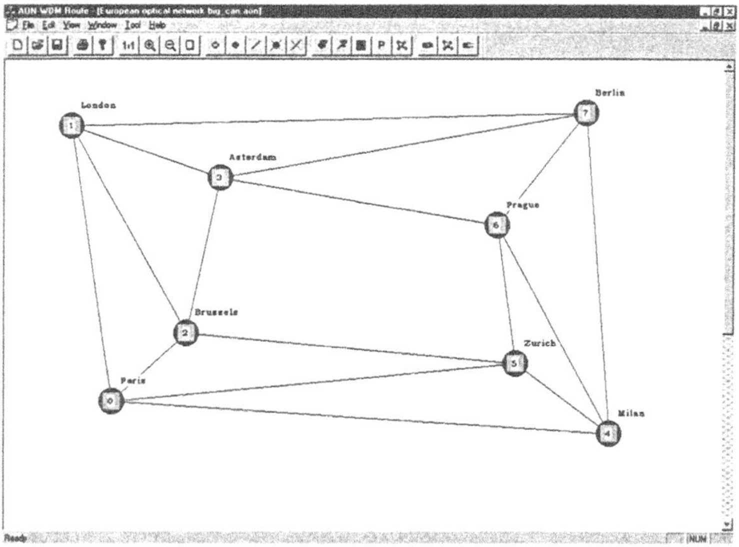

Fig. 6 European transport network topology

Table 1

Node pair	Distance [km]	Demand [channels]
0-1	480	36
0-2	300	28
0-3	560	12
0-4	740	28
0-5	640	32
0-6	780	4
0-7	1160	40
1-2	320	12
1-3	360	20
1-4	1040	16
1-5	860	12
1-6	840	4
1-7	1040	36
2-3	260	24
2-4	740	8
2-5	540	4
2-6	580	4
2-7	880	16
3-4	820	8
3-5	620	8
3-6	480	4
3-7	700	32
4-5	220	24
4-6	500	4
4-7	900	36
5-6	320	4
5-7	740	44
6-7	420	8

The goal of the planning process was to determine the minimum number of fibres for each link, which allows a wavelength allocation solution. In this way, 1, 4, 8 or 16 wavelengths per fibre have been considered. The results are listed in Table 2.

Table 2

Link	Link load [channels]	Fibres/link		
		4 λ	8 λ	16 λ
0-1	36	18	12	8
0-2	80	40	23	15
0-4	28	14	10	4
0-5	36	17	11	7
1-2	40	18	11	7
1-3	24	11	10	4
1-7	36	14	9	6
2-3	104	48	33	19
2-5	48	19	14	10
3-6	20	10	8	3
3-7	88	40	26	15
4-5	48	17	10	7
4-6	12	7	4	2
4-7	36	15	9	5
5-6	52	21	11	9
6-7	52	23	13	7

The capacity utilisation is defined as the ratio $l/(kL)$, where l is the link load, k is the number of fibres and L is number of wavelengths per fibre.

Fig. 7 displays the capacity utilisation for each link of the 8-nodes European network. It turns out that the capacity utilisation decreases when the number of wavelengths per fibre gets higher. This is due to the increase in the number of edges in the neighbourhood graph since a higher amount of channels share the same fibre. Anyway, as it is shown in Fig. 8, the use of higher number of wavelengths allows a substantial decrease of the total amount of fibre.

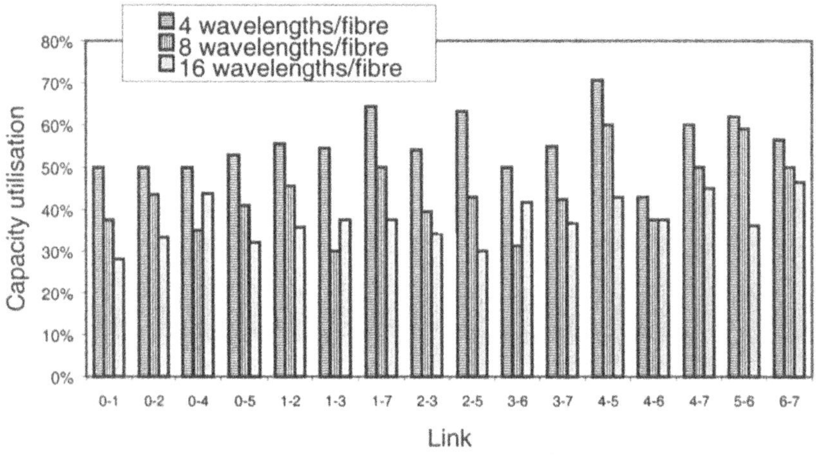

Fig. 7 Capacity utilisation

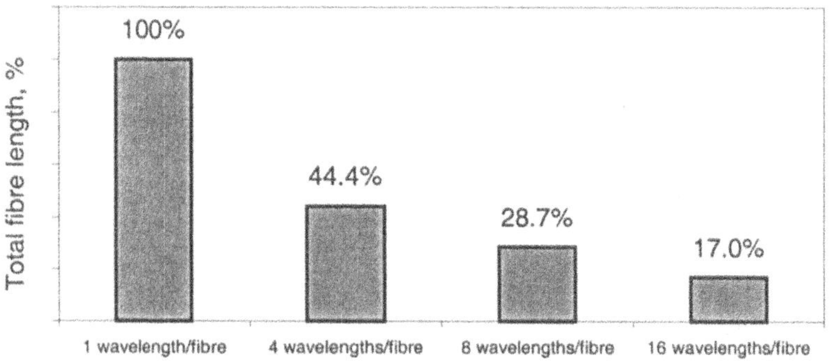

Fig. 8 Relative overall fibre length

4. CONCLUSION

The software tool described above performs a fast evaluation of an existing network or a new architecture under different load situations. Various "what if" scenarios can be tested allowing the planners to select the most suitable alternative. In the near future, this planning tool will be further developed to address key issues such as protection and quality of transmission.

ACKNOWLEDGEMENT

This work has been done within the COST Action 266 "Advanced infrastructure for photonic networks"

REFERENCES

[1] E. Lowe, "Current European WDM deployment trends," *IEEE Communications Magazine*, vol. 36, nr. 2, February 1998.

[2] A. McGuire and P. Bonenfant, "Standards: the blueprints for optical networking," *IEEE Communications Magazine*, vol. 36, nr. 2, February 1998.

[3] A. Kershenbaum, *Telecommunications network design algorithms*, McGraw Hill 1993.

[4] T.K. Tan and J.K. Pollard, "Determination of minimum number of wavelengths required for all-optical WDM networks using graph colouring" *Electronics letters*, vol. 31, nr. 22, February 1995.

[5] L. Besuchet, *Routage optimal et assignation des longueurs d'onde dans un réseau de communications optiques WDM*, Projet de diplôme 1060 EPF-Lausanne DE-TCOM 1999.

[6] A. Herz and D. de Wera, "Using Tabu search techniques for graph colouring" *Computing*, vol. 39, Springer-Verlag 1987.

[7] C. Bungarzeanu and A. Colarusso, "European optical transport network" *Proc. COST Workshop 94: Optical telecommunications*, Nice, April 17-21, 1994.

All-optical Signal Processing Devices for Photonic Networks: Towards Packet Switching

Hans-Peter Nolting
Heinrich-Hertz Institut für Nachrichtentechnik, Berlin GmBH, 10587 Berlin, Einsteinufer 35,
tel: +4930 31002 427, fax: +49 30 31002 241, nolting@hhi.de

Key words: **All-optical, signal processing devices, photonic networks, signal processing methods, terabit systems, terabit router, packet switching, OTDM-Systems,**

1. INTRODUCTION

The future internet demands for increase of the transmission capacity in access and core networks. Today telecom operators are installing massively WDM links to gratify the demand for capacity upgrades mainly for point-to-point connections. The link capacity will grow to a few Tbit/s per fiber. It is expected that optical time domain techniques (OTDM) will be used to increase the bit rate per WDM channel further beyond the highest available electronic processing speed. Today 10 Gbit/s-systems are commercial available, for the next months 40 Gbit/s systems are announced. With today's OTDM techniques an increase to 160 Gbit/s is feasible. It is expected that the bit rate for one wavelength channel may be increased to about 1 Tbit/s. The needs for the pulse sources, the demultiplexers, and the design of the fiber system will be discussed.

Today the optical networks are composed of point-to-point links with electronical switching and routing functions in the network nodes (SDH-,

ATM-hierarchy). The evolution of the network will go to more advanced nodes, where the signal will stay in the optical domain and will be switched, routed and processed by all-optical signal processing methods. For complex processing functions a hybrid solution may be advantageous, where extensive calculations will be done in the electrical domain using the superior performance of modern CPUs. But for some selected functions like all-optical high-speed switching (up to 100Gbit/s today), wavelength conversion and 3R-regeneration (up to 40 Gbit/s today) all-optical signal processing devices are investigated in system experiments and have shown their merits. These components are thought to be suitable for "moderate" bit rates like 10 to 40 Gbit/s system due to their performance in term of simplicity, advanced functionality, cost and reliability. The state of the art of these all-optical signal processing devices will be described and further evolution will be discussed.

Today monitoring and control of the network is done by the electronics (SDH, ATM). If the signal stays in the optical domain, for operation, control and maintenance new methods have to be developed based on optical signal processing methods. Optical sampling can be used for monitoring purpose of optical signals at very high bit rates. First results will be discussed.

The future internet will force a change from circuit switched networks to packet switched networks. If we use OTDM systems, than the well known bit-interleaving technique for the multiplexing / demultiplexing can not be used any more. Packets arriving at a lower bit rate at the node have to be compressed in time and bit shape and decompressed at the other end. Here some modelling results shall show the main problems to overcome. To reduce the processing time of the packet at a node the concept of "switching on the flight" will be studied. Assuming IP with label switching, the label has to be separated from the packet, read, analysed, a new label computed, the routing performed and the new label written to the packet. The label separation and label adding function will be done with all-optical processing functions. The packet itself should stay in the optical domain. For this a fiber memory with a 3R-regenerating function will be used. The label will be processed – in a first step - with a high-speed electronic unit.

Optical signal processing will play a key role in this scenario. In particular packet switching at ultra-high speed is a challenge. State of the art of optical signal processing in the bit rate range from 10 to 160 Gbit/s and an outlook into the future will be given, where bit rates of about 1 Tbit/s are announced.

2. OTDM-TRANSMISSION SYSTEMS

In Optical Time Division Multiplexing (OTDM) systems the multiplexing from a "low bit rate" (today 10 Gbit/s) is done by distributing short pulses from a ps-pulse source to a variety of n modulators (each working at 10 Gbit/s) and combining the individual delayed pulses to an interleaved bit stream of n*10 Gbit/s, as is shown in the figure.

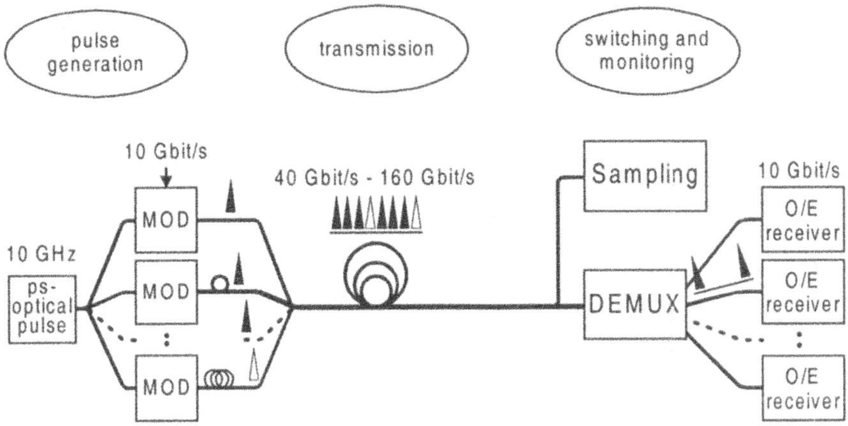

The critical device is the *ps-pulse source*. For the transmission of short pulses on standard fiber special care has to be taken, like dispersion management and / or polarisation mode dispersion compensation (PMD). Demultiplexen needs an all-optical switch, working with a switching window of the pulse width, but with a repetition rate of the basic bit rate. In order to switch at the right time slot, an optical *clock recovery* circuit gives the time information.

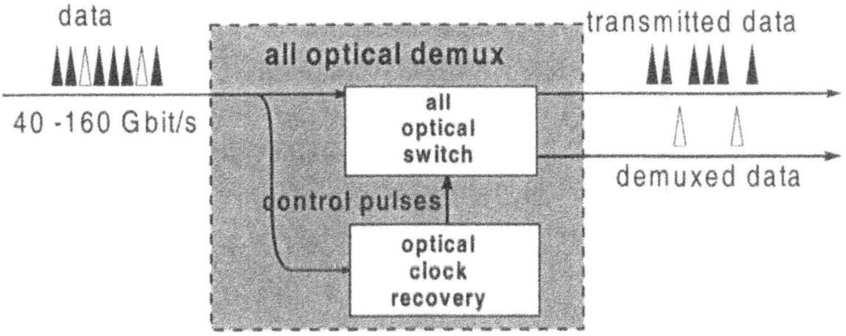

2.1 Pulse Source

For ultra-high speed transmission fs-pulses are necessary. One way to generate sub-pico second pulses is the so-called tunable mode locked laser (TMLL)[1]. Here a schematic of the pulse train is shown, summarising all important parameters of the TMLL in this plot[2].

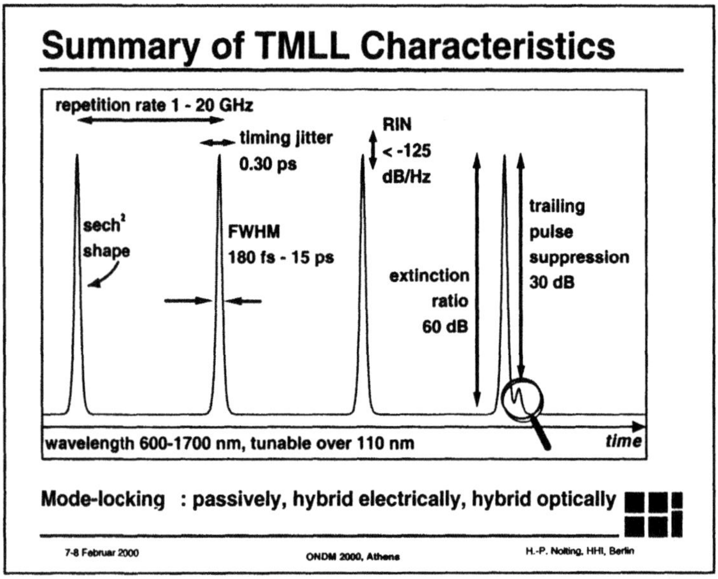

- The repetition rate of the TMLL can be tuned form 1 to 20 GHz by varying the cavity length.
- The pulses have a sech^2 shape for pulse widths down to 600 fs.
- The pulse width can be tuned from 180 fs to 15 ps by changing the grating and by switching from collimated to focussed projection onto the grating.
- The RMS timing jitter of the pulses is typically about 300 fs.
- The relative intensity noise is smaller than -125 dB per Hz.
- The extinction ratio is about 60 dB and the trailing pulse suppression is about 30 dB.

And finally, the centre wavelength can in principal be anywhere between 600 and 1700 nm depending on the type of laser diode used. The tuning range for a fixed diode is as large as 110 nm.

The saturable absorber is created by nitrogen ion implantation through one facet of the fabry perot laser diode.

Mode locking can be achieved passively, hybrid electrically and hybrid optically.

There is a variety of applications for ps-pulse sources in OTDM systems: (i) as transmitter, (ii) as clock recovery in 3R-regeneration (re-shaping, re-timing), (iii) synchronisation in demux and (iv) optical sampling.

Thus the ps-pulse source is the key device for OTDM-systems. The shortness of the pulses and the spectral purity determine the transmission capacity and the transmission length.

2.2 Dispersion Management

The propagation of pulses on fibers is limited by fiber loss, fiber dispersion, non-linearity and other effects. The invention of the Erbium-doped fiber amplifier (EDFA) has overcome the loss limit by periodic signal amplification on a long fiber link. The transmission of short pulses on standard fibers (SMF with a dispersion of 17ps/km/nm) leads to a broadening of the pulse width, which also limits the transmission length. The transmission length can be increased by compensating the accumulated dispersion by adding some fiber length with a negative dispersion (DCF fiber). For applications with very short pulses (terabit systems), which have a broad wavelength spectrum, or for WDM applications the accumulated dispersion has to be compensated in a wide wavelength region. In this case the dispersion compensating fibers have to compensate the dispersion itself and higher order dispersion.

The maximum propagation distance[3] of a transmission line composed of cascaded links of SMF, EDFA and compensating fiber is a function of the input power of the signal, as is shown in the figure. For low input power the limiting factor is the noise accumulation generated by the optical amplifier. For large input power the limiting factor is the combined influence of non linear effects.

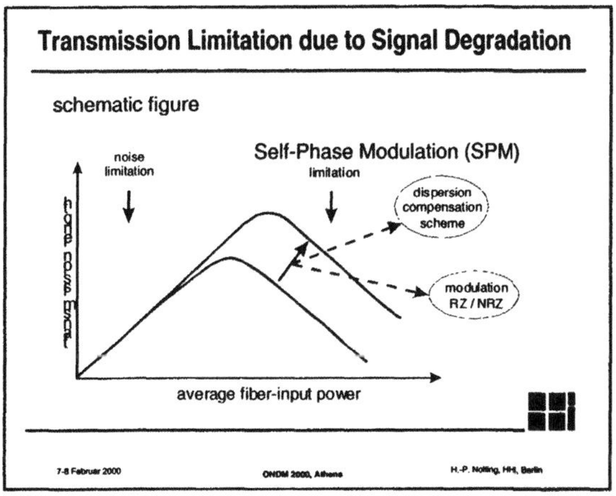

2.2.1 Terabit System Design

Modelling results on tera bit transmission systems are not available at this time. Some basic trends can be deduced from fundamental work at lower bit rates. In the first example experimental and simulation results are compared for transmission on dispersion compensated standard fiber links, while the two next are only theoretical studies of dispersion shifted fiber links at higher bit rates.

Recently four WDM-channels, each carrying a data rate of 40Gbit/s were transmitted over a 100km single mode fiber (SMF) link in RZ format[4]. The fiber link consisted of a 100km SMF span with an EDFA and a dispersion compensating fiber (DCF). The experiment was numerically simulated and the calculated results were in good agreement with experiment.

For the 4x40 Gbit/s experiment a DCF (labeled as DCF1) is used for which second order dispersion D is fully compensated but not the third order dispersion S. The results of the calculations depicted in the figure show that for such a compensation only up to three spans can be cascaded. If, however, DCF2 is used which compensates both D and S the penalty curve has the same shape but is 3dB lower. A similar curve results if dispersion is compensated in the receiver, e. g. by a fiber Bragg grating.

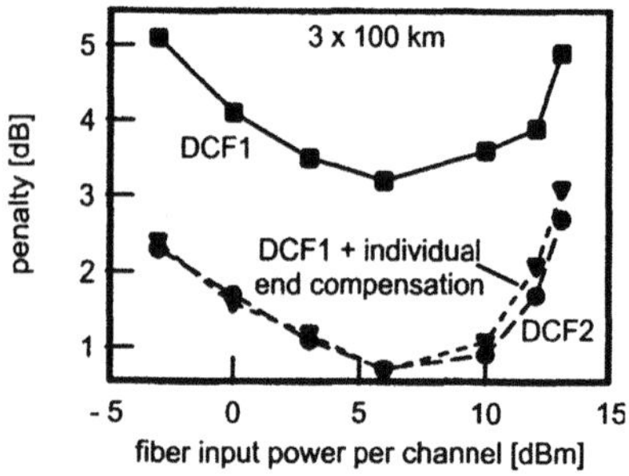

Next the numerically investigation of the cases of 8x40 Gbit/s, 8x80 Gbit/s and 8x160 Gbit/s transmission over 5x100km of SMF and 100km amplifier spacing are shown. In order to compare the results with the 4x40 Gbit/s experiment pulse widths of 4ps, 2ps and 1ps and channel spacings of 2nm, 4nm and 8nm are taken for 40Gbit/s, 80Gbit/s and 160Gbit/s, respectively. Each span is assumed to be completely compensated by DCF2. An important result of the calculation is that four-wave mixing and cross phase modulation effects are negligible for the dispersion map and channel spacings considered.

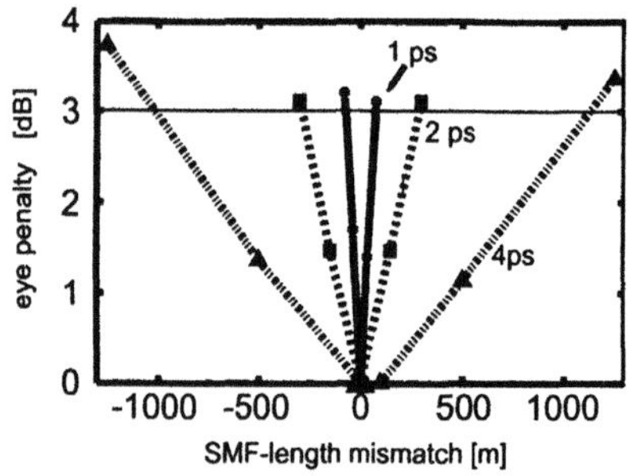

Further an evaluation of the stability of transmission with respect to small changes in the degree of dispersion compensation has been done, by deviating the SMF length from the value of complete compensation. The SMF-length tolerance is due to the increased broadening of the pulses for deviation of total dispersion from zero. The results of the calculations are shown in the figure. The optimum input power is 10dBm per channel. In this figure the eye closure penalty for deviations from the ideal fiber length is depicted. In order to keep transmission penalty below the 3dB penalty an SMF length tolerance range of 1.1km, 0.28km and 0.07km is necessary for the 4ps, 2ps and 1ps pulse width, respectively. Length tolerances well below 1km are difficult to realize and a way must be found to increase the numbers listed above. Since for non-soliton transmission pulse broadening by dispersion depends inversely on the square of the pulse width the above length tolerance are increased by a factor of 4 if pulse widths of 8ps, 4ps and 2ps are used for 40Gbit/s, 80Gbit/s and 160Gbit/s, respectively.

Kazuro Kikuchi[5] has discussed the design of an ultra-high speed optical transmission system based on a dispersion-shifted fiber. It is shown that the distance for dispersion compensation / amplifier section is decreasing for increasing bit rate.

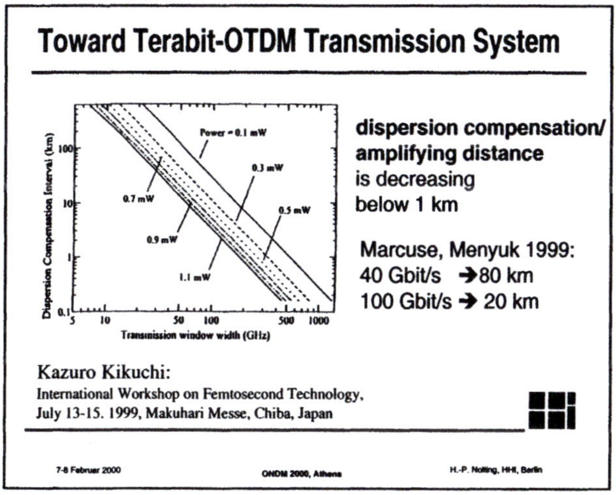

Marcuse and Menyuk have investigated theoretically the tolerances of a 100 Gbit/s transmission system for three different formats: non return to zero (NRZ), return to zero (RZ) and the dispersion-managed soliton (DMS) propagation. It is shown, that the amplifier span has to be reduced to 20 km compared to 80 km for 40 Gbit/s systems. Further the tolerances for power and dispersion variations are shown to decrease with increasing transmission length and bit rate. These results are in agreement with the experimental and theoretical results from HHI, given above.

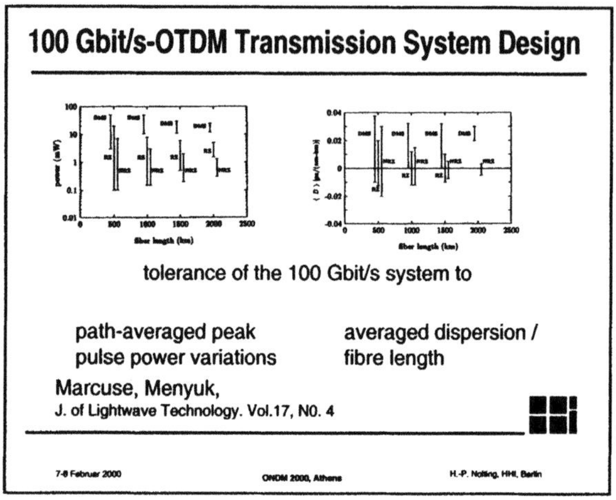

From this we can learn that the tolerances of the operation windows will become even more tight, if we go from 160 Gbit/s to much higher values like 1000Gbit/s. It looks like that the development of new fibers and new dispersion management concepts are necessary.

2.3 System Experiments

Some of the actual hero system experiments, aiming for highest transmission capacity on a single fiber, are shown in the table:

laboratory	WDM	OTDM	basic bit rate (electronic)	capacity [Gbit/s]
HHI[6], no transmission	8	8	10	640
Siemens[7]	80	1	40	3200
Lucent/BT[8]	1	8/16	20/10	160
NTT[9]	19	16	10	3040

At HHI fundamental investigations concerning the data format, the method of dispersion compensation, the amplifier span and other parameter are investigated and the results are shown in the following table.

	year	capacity [Gbit/s	description
HHI[10]	1997	40	150 km single span comparison of post- and precompensation comparison with TWF and DSF
HHI[11]	1998	40	430 km 4 spans mid-span spectral inversion
HH[12]	1998	4x40	100 km single span
HH[13]	1999	40	430 km 4 spans comparison of RZ and NRZ modulation format
HH[14]	1999	80	106 km 2 spans mid-span spectral inversion

2.4 Optical sampling for monitoring of signals

Network operators need information on the status of their network elements to have control over their network. For a variety of applications it is necessary to guarantee the quality of the service (QoS). One important parameter of a transmission line is the bit error rate (BER) of the link. In an OTDM-system the basic bit rate (electronic) is lower than the (optical) bit rate on the fiber. Therefore optical methods for measuring are necessary.

Optical sampling[15] can provide some insight in the shape of the signals and we hope to extend such a measurement setup to investigate the Q-factor of the system, which is correlated with the BER curves.

The principle of optical sampling[16] is shown in the figure. For a periodic signal it is possible to select information on the height of the amplitudes on a repetition rate, which is much lower than the bit rate of the signals.

To get a measure for the amplitude of the signal at a time a logical AND function has to be used between very short pulses as a control signal and the data.

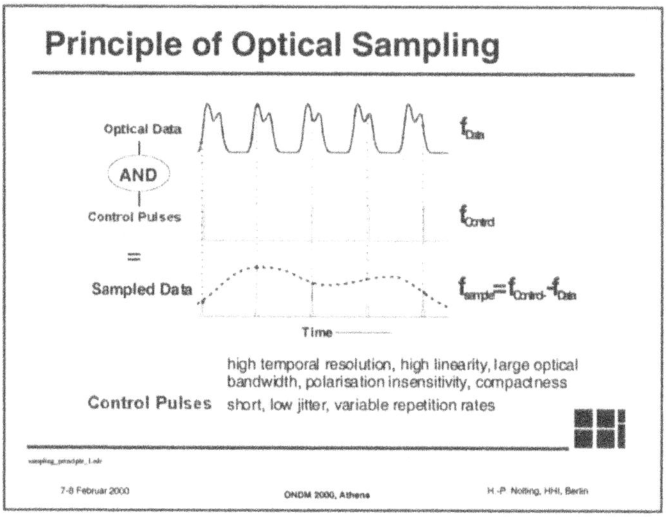

For this purpose the four wave mixing process in the gain-transparent switch between control and signal can be used, which generates a new signal that is proportional to the desired amplitude and can be filtered out.

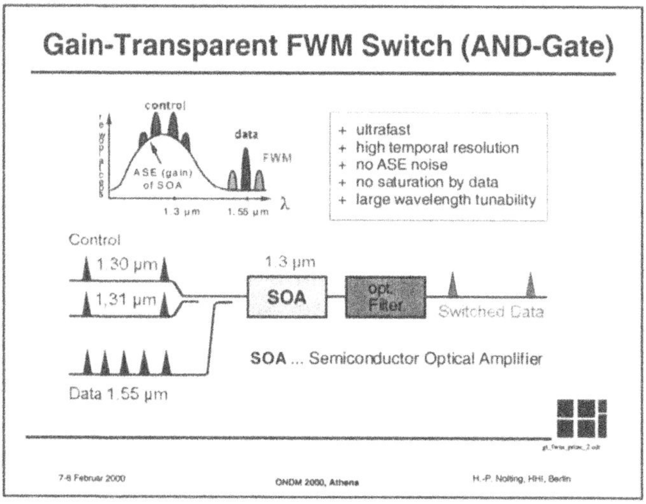

As is shown in the figure, two pumps around the wavelength 1,3 µm as control are necessary. The control pulses have to be as short as possible, a low jitter and the repetition rate has to be variable. The switch needs a high linearity, high temporal resolution, large optical bandwidth and polarisation insensitivity. This is fulfilled by the gain-transparent switch.

As first example the shape of a 160Gbit/s optical sampled data stream[16] is shown in the figure. This can not be done with pure electronic methods.

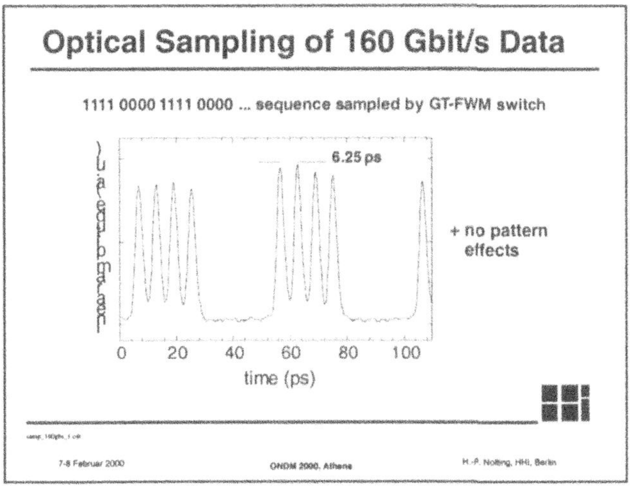

3. BASIC OPTICAL SIGNAL PROCESSING DEVICES

We use two lines of basic optical signal processing devices, both are based on semiconductor laser (amplifier) material. The advantages of this choice are (i) short device length (compared to fiber), (ii) robust and stable operation for integrated devices, (iii) cheap to fabricate and they offer (iv) gain .

For all-optical switching non coherent effects are used. Thus for data input and control input we need always two different wavelengths.

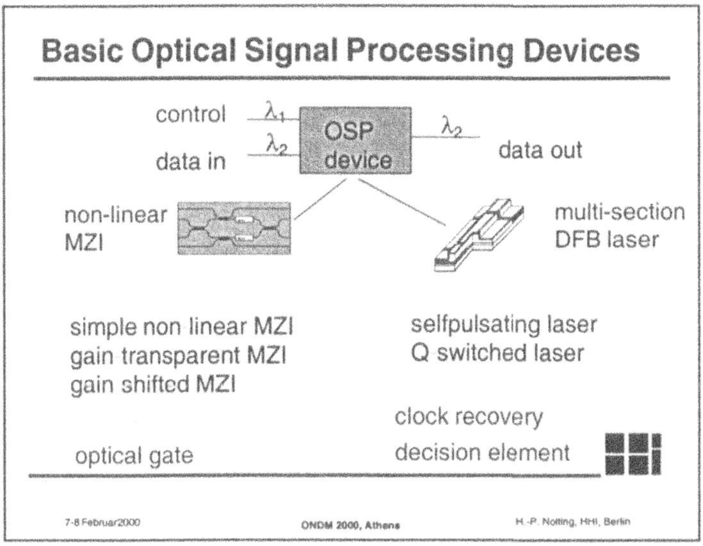

As optical gate the non-linear Mach-Zehnder Interferometer (MZI) with semiconductor optical amplifier (SOA) is used. The switching behaviour can be shaped by using different band gap material for the SOA (simple non-linear MZI, gain-transparent MZI, gain-shifted MZI). This component as a single pass device has a very high speed potential compared to laser devices, which are multi-pass elements.

For clock recovery a multi-section DFB laser operating as a selfpulsating laser can be used. By introducing data to the free running laser, the output signal can be locked to the incoming data stream. Basically the same device, but optimised for switching behaviour can be used as a decision element.

4. 3R-REGENERATION

Data transmitted over large distances show signal degradation due to the fiber non-linearities and dispersion effects. The passing of a network node in the optical domain will add further noise and crosstalk accumulation due to switching and wavelength conversion in the node. All-optical 3R signal regeneration (**R**e-amplification, **R**e-timing, **R**e-shaping) is a key function needed for scalable all-optical networks. The main functional blocks are the optical clock recovery and the decision element.

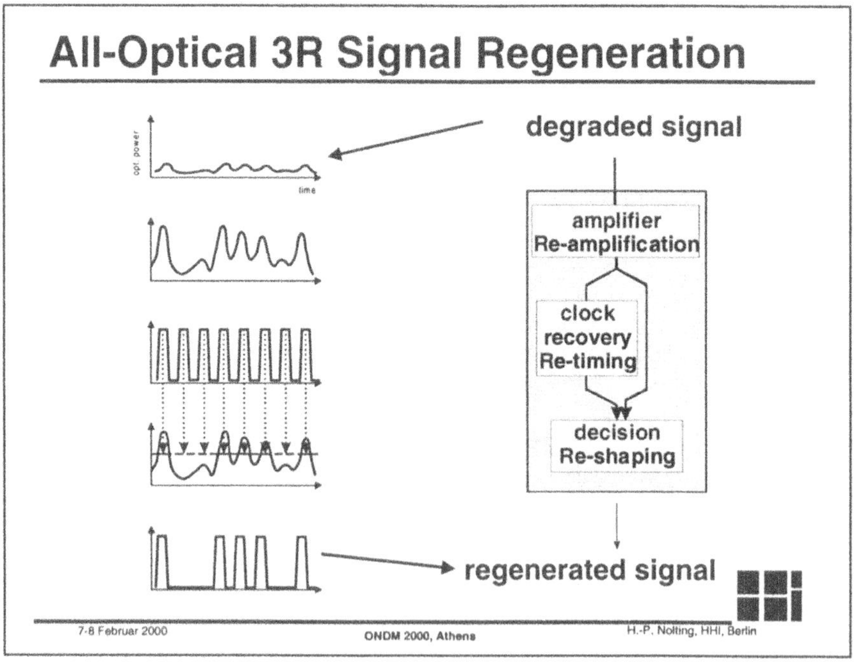

4.1 Non-linear Mach-Zehnder Interferometer as basic all-optical Switch

Interferometric switches incorporating semiconductor optical amplifiers (SOAs) have been widely investigated for high-speed all-optical signal processing. In these switches, an optical control signal depletes the carriers in the SOAs. The carrier depletion causes a gain change in the SOA which is accompanied by a change of refractive index. The latter leads to a phase change of the data signal and is used for interferometric switching. The wavelength of the control signal (blue) and the input signal (red) have to be different to avoid coherent effects.

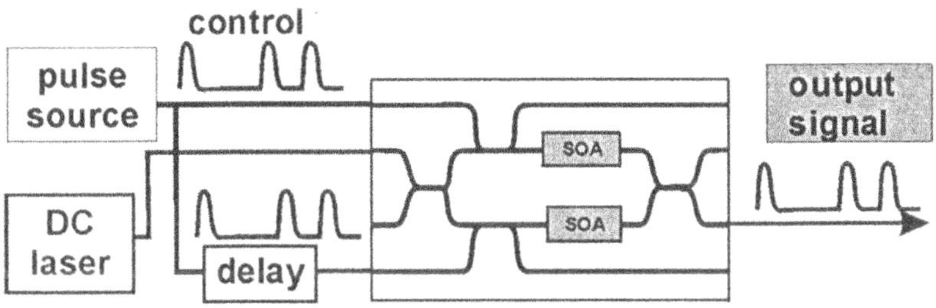

The device can be used for different applications:

- as wavelength converter with a constant amplitude of the input light
- as optical gate to led pass optical signals (red) on the input port to the output port (not shown in figure)
- in combination with a clocked signal this gate can be used as switch for a 3R-regenerator

We prefer an interferometer with two control inputs, to use push-pull operation with a short delay between both control signals (blue). This architecture can operate at higher speed than the single control interferometer.

The MZ-interferometer with long SOAs has a speed potential of about a few 100 Gbit/s.

The switching behaviour can be shaped by using different band gap material for the SOA:

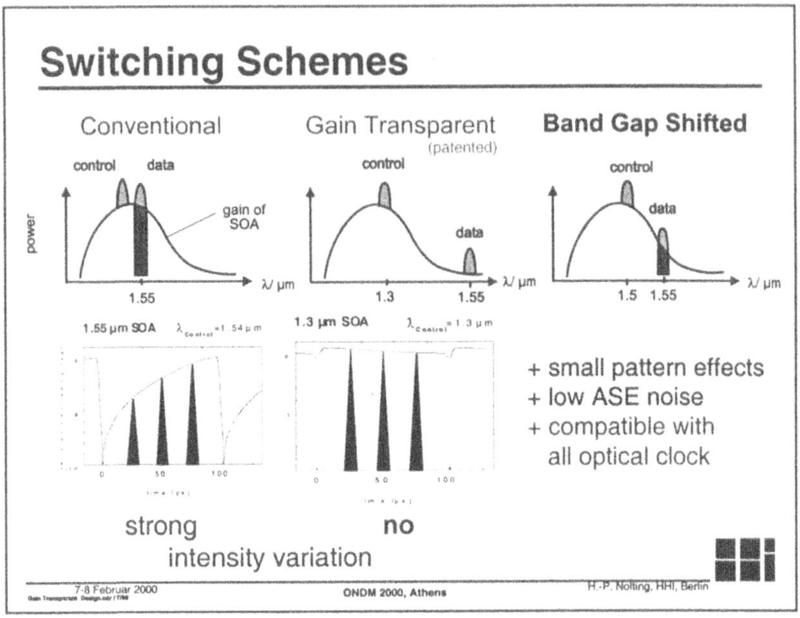

- *simple non-linear MZI,*[17] data and control pulses are both in the maximum of the gain curve. The resulting switching characteristic shows a strong pattern effect for the passed signals due to gain saturation effects
- *gain-transparent MZI*[18], only the control pulses are in the gain maximum. The data pulses are far away from the gain maximum and will be switched only by phase changes. The amplitudes of the signal are nearly constant.
- *gain-shifted MZI*, this is a new concept, where the data wavelength is placed at the border of the gain characteristic. This allows to operate this switch in combination with wavelength converters, which is necessary in some system applications. One example is the 3R-regenerator, where input and output wavelength often have to be identical, but a wavelength conversion is accompanied with a one stage gate function. Here an additional wavelength conversion is necessary.

4.1.1 Tbit/s demultiplexing with LOTOS

A new type of very fast switching device has been reported. This switch uses low temperature grown InGaAs/InAlAs[19] multiple quantum wells in a free space arranged Fabry-Perot resonator. The module achieves a response time of 390 fs with a contrast ratio of 25 in a fiber-based experiment. Using the modules, all-optical demultiplexing is demonstrated for signal pulses corresponding to 1.05 Tbit/s, and the demultiplexed optical pulses have been clearly detected.

This ultra-fast switching effect is accompanied with high losses.

4.2 Selfpulsating Laser as Clock Recovery

The selfpulsating laser[20] is a three section DFB laser, which has been developed specially for the clock recovery function. The device has a gain section, a phase shifter section and a Bragg reflector section. Here the selfpulsation is generated by the feedback from the reflected wave of the Bragg reflector into the gain section under control of the phase section. The device is operated with three DC currents, which control the following functions:

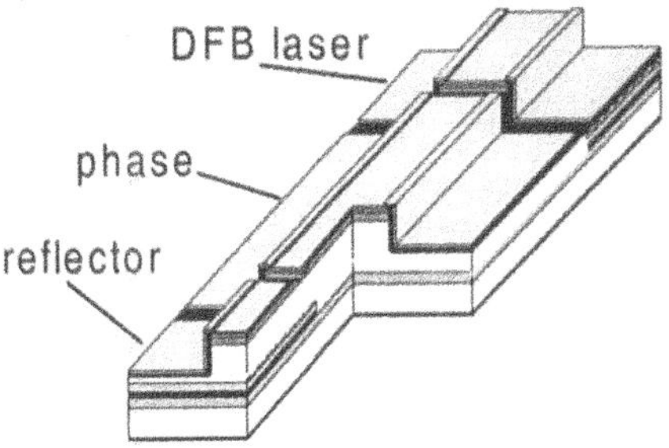

- phase current: selfpulsation on / off
- gain current: tuning of the selfpulsation frequency[21] with a continuous tuning range from 6 to 40 Gbit/s[22]
- Bragg current: tuning of the wavelength (a few nm)

Injection of a data stream (both input wavelengths have to be different, to avoid coherent effects) into the free running selfpulsating laser locks the output signal to the data stream.

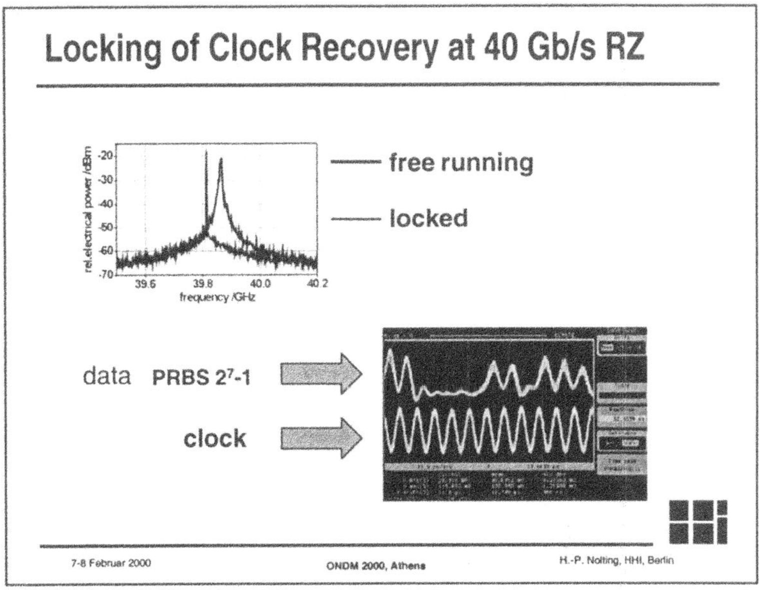

4.3 Q-switched laser as decision element

Bistable Q-switching has been demonstrated static and up to 10 Gbit/s for a specially optimized 3 section DFB laser device, which can be used as a decision element in all-optical signal processing application. This device can easily be integrated monolithically together with the clock recovery circuit. The static switching characteristic is shown.

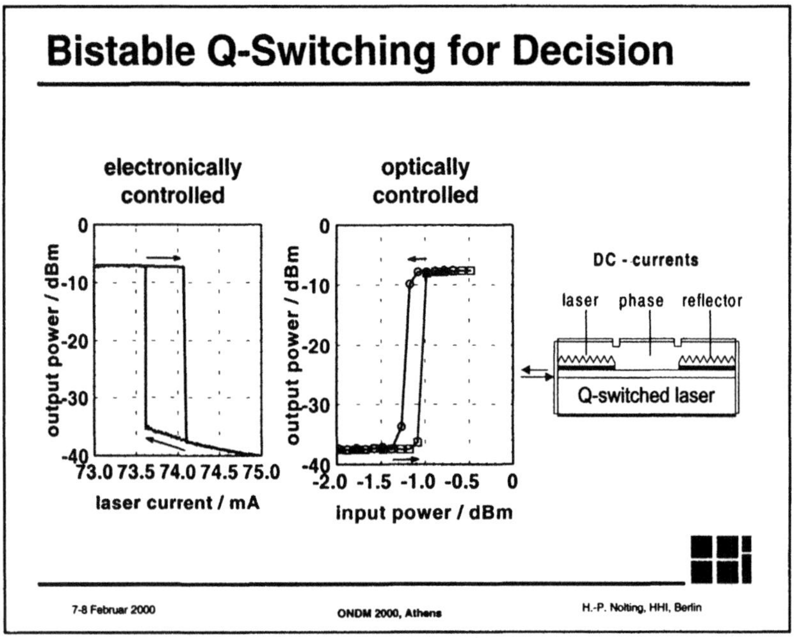

4.4 3R-Regenerator System Experiments

In cooperation of Alcatel and HHI a 10 Gbit/s transmission experiment has been successfully conducted with a 3R-regenerator composed of two MZ-interferometers (input wavelength equal to output wavelength) and a selfpulsating laser as clock recovery. The basic setup and the BER measurement curves are shown.

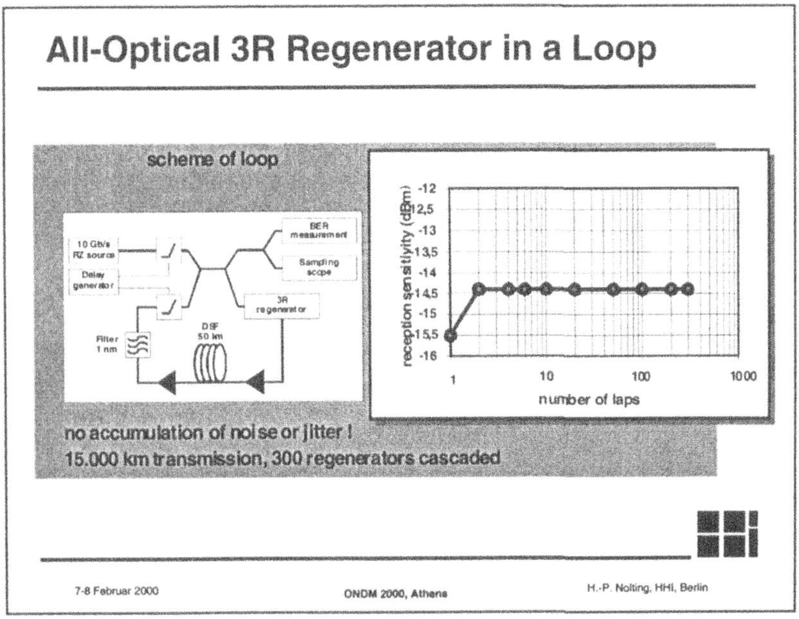

Further on a loop transmission experiment with a repeater length of 50 km length has been conducted. After the first lap a small penalty of about 1 dB has been observed. The next 300 laps show no additional penalty, indicating a successfully limiting of noise and jitter accumulation by this all optical 3R-regenerator.

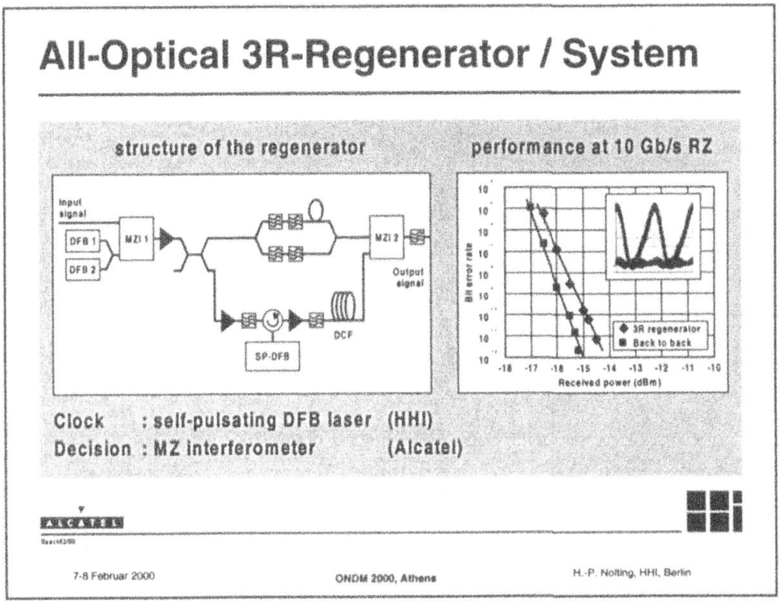

5. OPTICAL PACKET SWITCHING

Packet switching based on the Internet Protocol (IP) is a connectionless oriented traffic, basically invented for computer networks. The data are put into packets and sent randomly to the neighbours. All packets are built up of the data or payload and an additionally header, where the address is put in. Each node of the network has an address table of other neighbouring nodes.

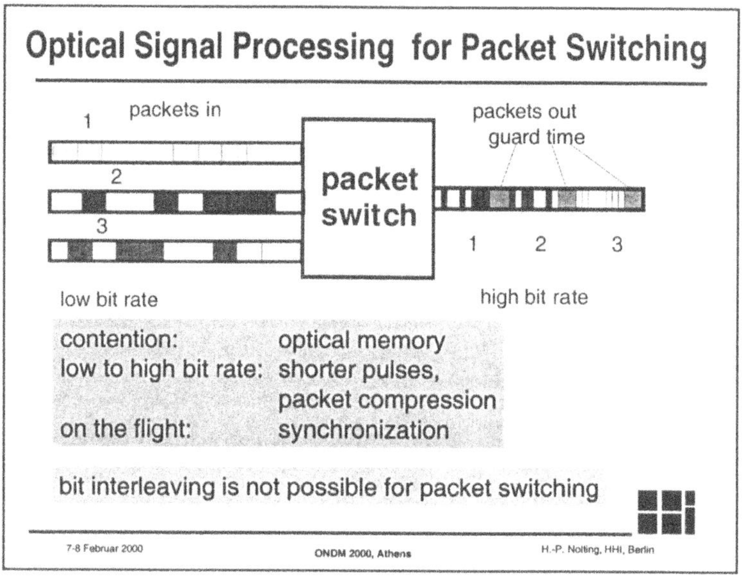

Based on this knowledge the packets are send to the next node, until it reaches its destination. Therefore the basic steps in a node are the following: storing a packet, identifying the address in the header, determining the output port of the node, writing a new header, sending the packet. All packets arriving in a node are queued until the processor has time to process the package. This signal processing can easily be done by today's computer, but it is a slow process, which increases the delay time for a packet propagating through the network.

Optical signal processing offers the chance for much faster processing time, especially if the packet is staying in the optical domain "on the flight", while the header is processed and a very fast switch is activated to send the packet to the next node. The non-linear MZ-Interferometer switch can be used as fast packet switch.

One concept is label switching in IP, where the path through a domain is planned at the border nodes (blue) and the packet gets an additional label, which is processed by optical signal processing methods in the inner nodes (red) of the domain. The introducing of a label domain reduces the length of the address table and the processing time and includes some other advantages.

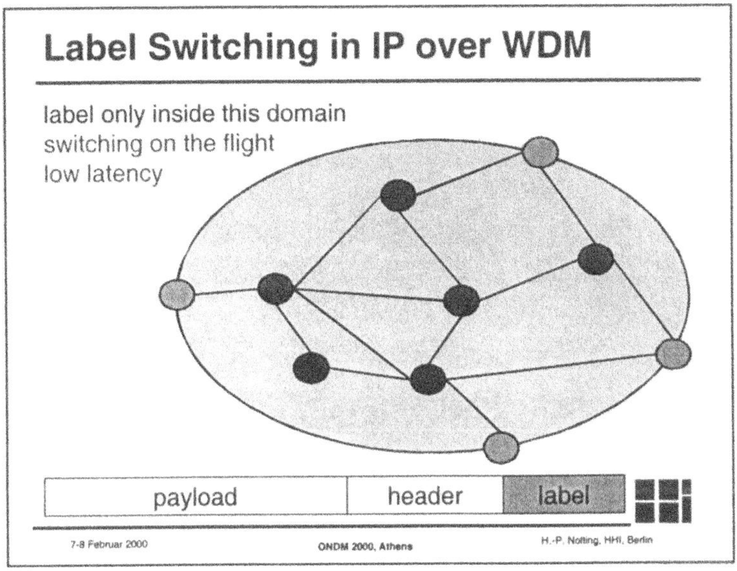

5.1 Optical Memory

In any case an optical memory is necessary for storing the packets during processing time. A well known solution is a fiber-ring memory with a 3R-regeneration to stabilise the signal shape. Again fast switching to open and shut the ring memory can be done by a MZI-switch. Synchronisation between two data streams is necessary. For this purpose switched delay lines have been investigated by several authors.

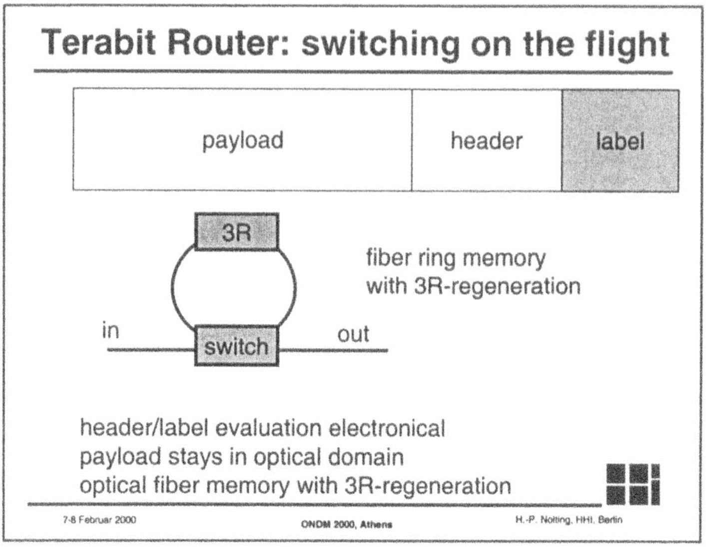

5.2 Fast locking clock recovery for packet switching

Packet switching is a burst mode of operation. Consecutive following packets are not synchronised as is usual in SDH. Therefore fast locking clock recovery circuits are needed. Here the selfpulsation multi-section DFB

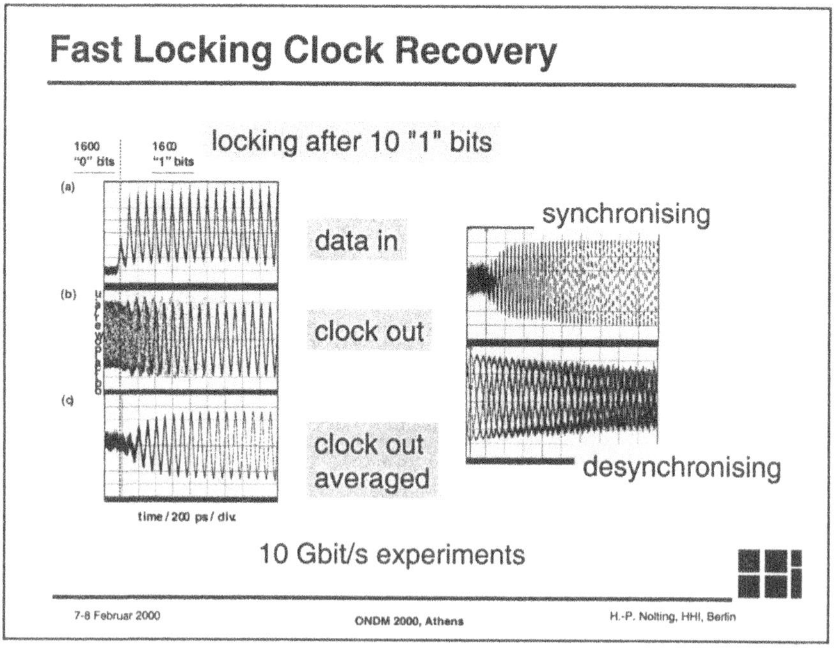

laser has its superior performance[23]. In experiments at 10 Gbit/s the synchronisation (locking) time has been investigated to a number of 10 "1" bits following a queue of 1600 "0" bits, representing the guard time of packets (see left side of figure "Fast Locking Clock Recovery"). This is a factor of 1000 shorter than is usual in today's electronic PLL circuits.

The number of "0" bits until the clock is desynchronised is in the order of a few hundred (see right side of figure). This "fast locking" function together with the "long memory" effect makes the selfpulsating DFB laser well suitable for all-optical packet switching applications. Together with fast optical gating devices a 3R-regenerator for packet switching can be built.

5.3 Packet Compression and Decompression

In OTDM transmission systems, the often used bit-interleaving technique for the multiplexing / demultiplexing can not be used any more for packets switching. Packets arriving at a lower bit rate at the node have to be compressed in time and bit shape and decompressed at the other end.

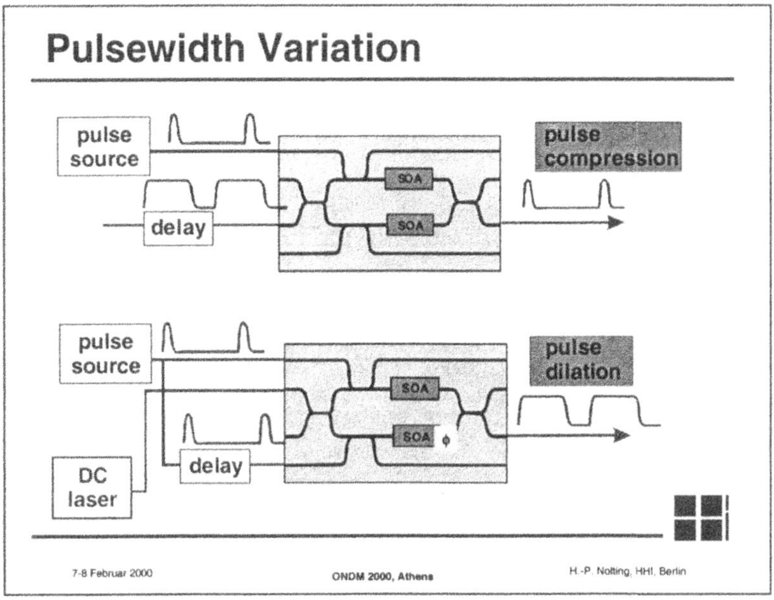

Non-linear MZ-Interferometer can be used for changing the width of signals. The MZ-interferometer acts as a logical and-function. A pulse source with short pulses synchronised to the data stream cuts a short pulse shape out of an incoming "1" bit. Thus the shape is compressed, but the time position is unchanged.

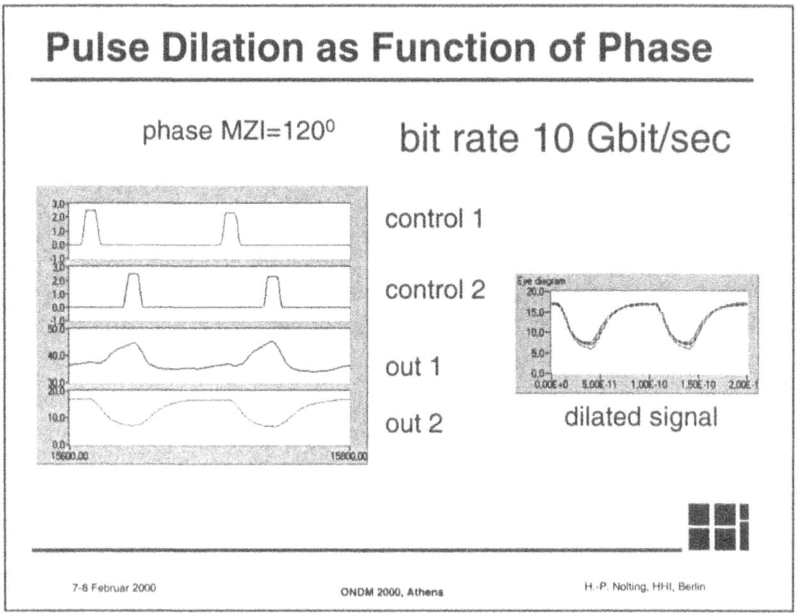

To increase the width of short incoming pulses they can be introduced at the control input of the MZ-interferometer in push-pull operation. The introduced delay determines the new pulse width. Here the coherent interaction of both pulses is used to make a flat output signal of the desired width. To control the shape an additional phase shifter is necessary. Some modeling results are shown.

This shaping of the pulses has to be done before and after the packet compression / decompression process. In the following we assume that the pulses are short enough for the higher bit rate.

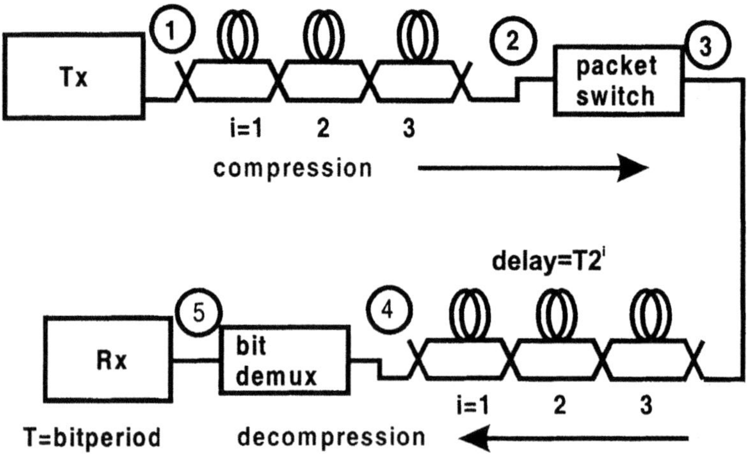

A method to compress and to uncompress a packet has been described by Prucnel[24]. As is shown in the figure, a passive network of n asymmetric Mach-Zehnder-Interferometer with a sequence of special dimensioned delays can be used to build up a compressed copy (at point 2) of the incoming packet (at point 1) of 2^n bits. This packet has to be cut out (at point 3) by a fast optical switch from the other not usable copies of this device. The basic function of this device is shown in the following modelling results with a compression factor of 10 in the following figure. To identify correctly all bits in a packet, the incoming packet is modulated in the height of the amplitude (point 1). Please notice the very short pulses. At point 2 a sequence of copies is shown. Only one copy is the complete packet and has to be switched by the packet switch (3).

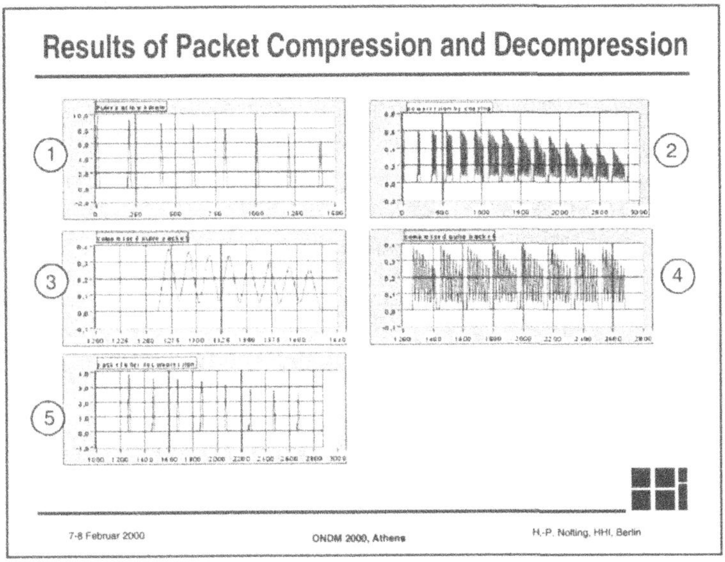

For the reverse process (decompression, dilation) the same sequence of passive MZIs can be used, as is shown in the lower part of both figures. At point 4 a sequence of copies of the original incoming short packet is shown. The bit demux has to switch out exactly one bit of the sequence at every period time T. The resulting packet is shown at point 5.

This concept has the advantage to compress and decompress, as example a packet of 2^{10}=1024 bits with 10 MZIs in a sequence. The disadvantage is the fact, that the number of bits in a packet is limited to the quotient of the two bit rates. For a realistic scenario of 10 to 80 Gbit/s this is not acceptable.

A much more realistic concept is to use a ring memory[25] to collect all bits of a packet without any limitation of the packet length. This concept has the disadvantage that the bits are scrambled in a special order and therefore the compression and decompression has to be done in pairs of devices. But this is always the case in practice.

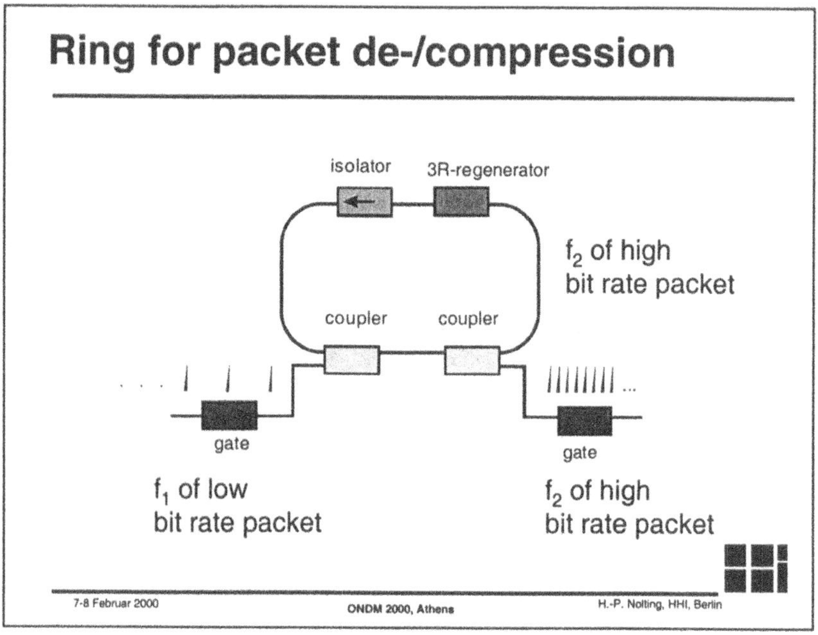

The scheme is based on the idea of self bit-interleaving (sometimes called rational harmonic mode locking[25]). Again we assume that the pulses are short enough for a bit-interleaving procedure, as is shown in both figures. To achieve rate conversion from low bit rate f_1 to high bit rate f_2 (as example take 10 to 40 Gbit/s) the packet is divided into n slices of bit streams (one common color in figure) with the length of m bits. Here n is the ratio of both frequencies ($n=f_2/f_1$). m is the lowest number that fits $N \leq n*m-1$, where N is the maximum number of bits in the whole packet. In the example shown in the figure we have less or equal than 31 bits and n=4 slices with a length of m=8 time slots. Please notice, that the last slice has only m-1=7 bits.

The ring length has to be adjusted to a length of exact n*m-1 time slots at the higher bit rate f_2.

That means that the first slice of the low bit rate packet (m=8 bits in our example, yellow colored as is shown in the figure) enter the ring by the left coupler and start to propagate in the ring memory. The next slice of bits

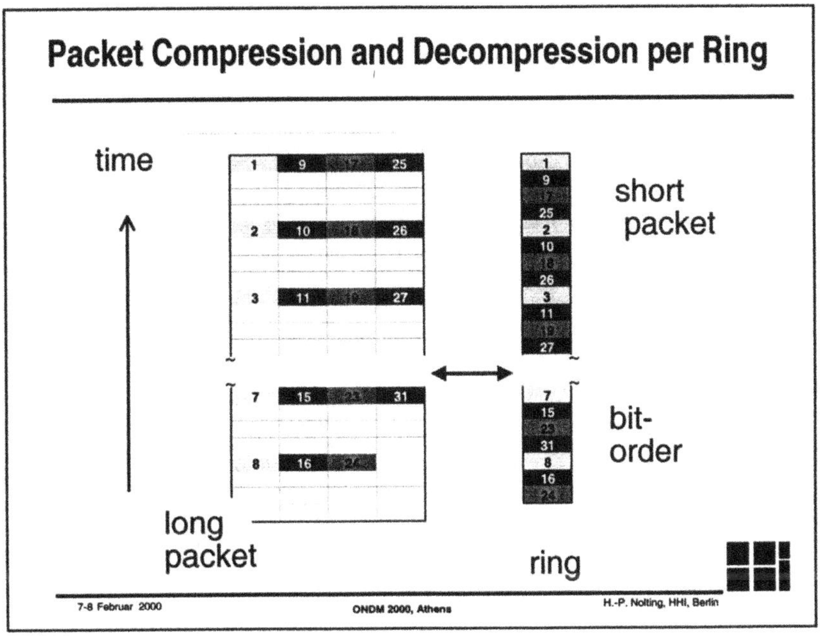

(number 9 to 14, green colored) meet a (short) time slot behind the bit numbers 1 to 8. This behavior is described by the name self bit-interleaving process. The next slices (colored red and blue) are also slipped by one (short) time slot. All bits will find a well defined place in the compressed package. The gate at the right side controls the passing of the complete package.

Decompression can be done by inverting the self bit-interleaving process. The same ring memory architecture can be used. The gate at the right side has to open the output arm for a short time slot with the period of the low bit rate.

6. SUMMARY

The state of the art of the application of optical signal processing devices and subsystems in present OTDM-systems in the bit rate region 10 to 160 Gbit/s have been highlighted. Some of the requirements and demands for future 1 Tbit/s systems and for future all optical packet switching at ultra-high speed have been discussed. All optical signal processing methods and application will play a key role. New network concepts and the speed enhancement of semiconductor based materials or the development of new materials for all optical switching is a challenge.

---

[1] **Tunable mode locked laser as pulse source**

R.Ludwig, A. Ehrhardt Turn-key-ready wavelength-, repetition rate- and pulsewidth-tunable femtosecond hybrid modelocked semiconductor laser, Electronics Letters, Vol.31, 6th July 1995, No.14, pp. 1165-1167.

R. Ludwig, S. Diez, A. Ehrhardt, L. Küller, W. Pieper, H. G. Weber, A Tunable Femtosecond Modelocked Semiconductor Laser for Applications in OTDM-Systems, *IEICE Trans. Electron., Vol. E81-C, No.2 (1998), pp. 140-145.*

[2] C. Schmidt, E. Dietrich, S. Diet, H. J. Ehrke, U. Feiste, L. Küller, R. Ludwig, and H.-G. Weber, "Mode-locked semiconductor lasers and their applications for optical signal procesing", Conf. On Lasers and Electro-Optics (CLEO) Tech. Dig. P. 347-349, May 1999, paper CthA3

[3] H. G. Weber, D. Breuer20, U. Feiste, I. Koltchanov20, R. Ludwig and C. M. Weinert, High-Speed Transmission over Standard Fiber, *Proc. 25th Europ. Conf. Opt. Commun. 1, September 1999, pp. 196-199.*

[4] W. Pieper, R. Ludwig, C. M. Weinert, B. Kuhlow, G. Przyrembel, M. Ferstl, E. Pawlowski, and H. G. Weber, „4-channel x 40 Gb/s unrepeatered OTDM transmission over 100km standard fiber", IEEE Photon. Technol. Lett, vol. 10, pp. 451-453, 1998

[5] Kazuro Kikuchi, "Ultra-high Speed Optimal Transmission System Design" The 6th International Workshop on Femtosecond Technology, July 13-15. 1999, Makuhari Messe, Chiba, Japan

[6] S.Diez, R.Ludwig, H.G.Weber: "All-optical switch for TDM and WDM/TDM systems demonstrated in a 640Gbit/s demultiplexing experiment", *Electronics Letters, Vol.34, No.8, (1998) pp803-805*

[7] J.-P. Elbers, C. Scheerer, A. Färbert, C. Glingener, A. Schöpflin, E. Gottwald, G. Fischer, "3,4 Tbit/s (80x40 Gbit/s) Bidirectional DWDM/ETDM Transmission", ECOC 99, September 26-30, 1999 Nice, France, Post-Deadline paper PD2-5, p 32-33

[8] B. Mikkelsen, G. Raybom, R.-J. Essiambre, K. Dreyer, Y. Su, L. E. Nelson, J. E. Johnson, G. Shtengel, A. Bond, D. G. Moodie, A. E. Ellis, "160 Gbit/s singel-channel transmission over 300 km nonzero-dispersion fiber with semiconductor based transmitter and demultiplexer", ECOC 99, Septembre 26-30, 1999, Nice, France, PD2-3, p32-33

[9] S. Kawanishi, H. Takara, K. Uchiyama, I. Shake, K. Mori, "3 Tbit/s (160x19 ch) OTDM/WDM transmission experiment", *Tech. Digest of OFC'99, PD1 (1999)*

[10] R.Ludwig, W.Pieper, H.G.Weber, D.Breuer, K.Petermann, F. Küppers, A.Mattheus, "Unrepeatered 40 Gbit/s RZ single channel transmission over 150 km of standard fiber at 1.55 μm", *Electronics Letters , Vol. 33, 2nd January 1997, No.1, pp. 76-77*

[11] U.Feiste, R.Ludwig, E.Dietrich, S.Diez, H.J.Ehrke, D.Razic, H.G.Weber, "40 Gbit/s transmission over 434 km Standard-Fiber Using Polarisation Independent Mid-Span Spectral Inversion", *Electronics Letters, Vol. 34, 15th October 1998, pp. 2044-2045*

[12] W. Pieper, R. Ludwig, C. M. Weinert, B. Kuhlow, G. Przyrembel, M. Ferstl, E. Pawlowski, H. G. Weber, "4-channel x 40 Gbit/s unrepeatered OTDM-transmission over 100 km standard fiber", *IEEE Photon. Technol. Lett., Vol. 10, No. 3, March 1998, pp. 451-453*

[13] R. Ludwig, U. Feiste, E. Dietrich, H.-G. Weber, D. Breuer, M. Martin, F. Küppers, "Experimental comparison of 40Gbit/s RZ and NRZ transmission over standard single mode fiber", *Electr. Letters, (1999), in press*

[14] U. Feiste, R. Ludwig, C. Schmidt, E. Dietrich, S. Diez, H.-J. Ehrke, E. Patzak, H.-G. Weber, T. Merker, "80 Gbit/s transmission over 106 km standard fiber using optical phase conjugation in a Sagnac interferometer" *IEEE Phot. Techn. Lett. Vol 11, (1999),pp 1063-1065*

[15] **Optical sampling**

S.Diez, C.Schmidt, D.Hoffmann, C.Bornholdt, B.Sartorius, H.G.Weber, L.Jiang, A.Krotkus , Simultaneous Sampling of Optical Pulse Intensities and Wavelengths by Four-Wave Mixing in a Semiconductor Optical Amplifier, *Applied Physics Letters, Volume 73, Issue 26, pp. 3821-3823, December 28, 1998.*

[16] S.Diez, R.Ludwig, C.Schmidt, U.Feiste, H.G.Weber, "160 Gbit/s optical sampling by a novel ultra-broadband switch based on Four-Wave Mixing in a Semiconductor Optical Amplifier", *Optical Fiber Communication Conference 1999, OFC '99, Technical Digest, post-deadline- paper PD38*

[17] **Conventional switch**

E. Jahn, N. Agrawal, W. Pieper, H.-J. Ehrke, D. Franke, H. G. Weber, W. Fürst, "Monolithical integrated nonlinear sagnac interferometer and its application as a 20 Gbit/s all-optical demultiplexer", *Integrated Photonics Research (IPR'96) (Boston,USA), April 1996, Techn. Dig. Vol. 6 paper ITuG3-1.*

N. Agrawal, E. Jahn, W. Pieper, H.-J. Ehrke, "Semiconductor Laser Amplifier Based Integrated Interferometers for Add/Drop Multiplexing in OTDM Sytems", *Fiber Integrated Optics, 10.-11. October 1996 (Berlin,Germany), SPIE Vol 2954, Proc. invited paper, pp. 166-177.*

[18] **Gain-Transparent switch**

S.Diez, R.Ludwig, H.G.Weber:, "All-optical switch for TDM and WDM/TDM systems demonstrated in a 640Gbit/s demultiplexing experiment", *Electronics Letters, Vol.34, No.8, (1998) pp803-805*

S.Diez, R.Ludwig, H.G.Weber , "Gain-Transparent SOA-Switch for High Bitrate Add/Drop Multiplexing", IEEE Photon. Technol. Lett., vol. 11, no. 1, pp. 60-62, 1999.

R. Ludwig, S. Diez, H. G. Weber:, "A novel all-optical switch for demultiplexing in OTDM-systems demonstrated in a 640 Gbit/s WDM/TDM experiment", *Optical Fiber Communication Conference 1998, OFC '98, postdeadline paper PD22 (1998)*

[19] H. Kobayashi, R. Takahashi, Y. Matsuoka and H. Iwamura, "1 Tbit/s demultiplexing using low temperature grown InGaAs/InAlAs multiple quantum wells", Elect. Letters, Vol. 34, No, 9, p. 908-910

[20] **Selfpulsating laser as clock recovery**

B. Sartorius, C. Bornholdt, O. Brox, H.J. Ehrke, D. Hoffmann, R. Ludwig, and M. Möhrle:
"System performance of an all-optical clock recovery module" ECOC'98, Madrid, Conf. Proc. pp.505-506, 1998.

A. Buxens, A.T. Clausen, H.N. Poulsen, K.S. Jepsen, K.E. Stubkjaer, C. Bornholdt, O. Brox, and B. Sartorius: 40 to 10 Gb/s demultiplexing using a self-pulsating DFB laser for clock recovery, ECOC'98, Madrid, Conf. Proc. pp.507-508, 1998.

B. Sartorius, C. Bornholdt, O. Brox, H.J. Ehrke, D. Hoffmann, R. Ludwig, and M. Möhrle:
"All-Optical Clock Recovery Module Based on a Self-Pulsating DFB Laser", Electron. Lett., 34, pp.1664-1665, 1998.

B. Sartorius, C. Bornholdt, O. Brox, M. Möhrle , P. Brindel, O. Leclerc and E. Desurvire :
"Analysis and Compression of Pulses Emitted from an All-Optical Clock Recovery Module"

Electron. Lett. 34, pp 2344-2345, 1998

[21] B. Sartorius, C. Bornholdt, O. Brox, H.J. Ehrke, D. Hoffmann, R. Ludwig, and M. Möhrle:"Bit-rate flexible all-optical clock recovery", OFC'99, San Diego, USA, paper FB1

[22] C. Bornholdt, B. Sartorius, M. Möhrle, "All-Optimal Clock Recovery at 40 Gbit/s", *ECOC 99, September 26-30, 1999, Nice France, Post-Deadline Paper PD3-5, p 54-55*

[23] S. Bauer, C. Bornholdt, O. Brox, D. Hoffmann, M. Möhrle, G. Sahin, B.Sartorius, S. Schelhase, B. Lavigne, and D. Chiaroni, " Ultrafast Locking Optical Clock for IP Packet Switching Applications", paper TuF5 at OFC 2000, Baltimore, USA

[24] P. P. Prucnal, S. W. Seo, Applied Optics, 35, pp. 3815-3818 (1996)

P. Toliver, K.-Li Deng, I. Glesk, P. R. Prucenal, "Simultaneos Optical Compression and Decompression of 100 Gbit/s OTDM Packets Using a Single TOAD and a Bi-directional Optical Delay Line Lattice, Photonics in Switching, Technical Digest, July 21-23. 1999. Santa Barbara, Ca, USA, paper PWB2-1/35

[25] Chongjin Xie, Peida Ye, "A novel scheme of optical packet-compression and –decompression for all-optical packet switching networks", *ECOC 99, 26-30 September 1999, Nice, France, p I-258-259*

Numerical Modeling of a High Repetition Rate Fiber Laser, Mode – Locked by External Optical Modulation of a Semiconductor Optical Amplifier

K. Zoiros, T. Stathopoulos, K. Vlachos, A. Hatziefremidis,
T. Houbavlis, T. Papakyriakopoulos and H. Avramopoulos
National Technical University of Athens Department of Electrical and Computer Engineering, Photonics Communications Research Laboratory, 9 Iroon Polytechniou Street, 157 73, Zografou, Athens, Greece, Tel : +30 1 772 2076, Fax : +30 1 7722077, Email : hav@cc.ece.ntua.gr

Key words: *Active mode - locking; High repetition rate ring laser; Semiconductor optical amplifier; External optical modulation*

Abstract: A theoretical study of a high repetition rate laser source based on a novel mode - locking technique is presented. This technique relies on the fast saturation and recovery of a semiconductor optical amplifier induced by an external optical pulse and has been used to obtain 4.3 ps pulses at 20 GHz. A numerical model of the fiber ring laser has been developed describing the mode - locking process in the laser oscillator and providing solutions for the steady - state mode - locked pulse profile. The critical parameters of the system are defined and analyzed and their impact on the formation of the mode - locked pulses is examined. The comparison between the theoretical results and the experimental data reveals very good agreement and has allowed the optimization of the performance of the system in terms of these parameters.

1. INTRODUCTION

Ultra-high speed multi-access networks are evolving rapidly and are being developed to satisfy the increasing bandwidth demand due to the massive use of Internet and multimedia. Optical signal sources capable of generating stable ultra - short pulse trains at high repetition rates and with a tuning range as wide as possible are key elements for these novel photonic networks that may combine wavelength division multiplexing (WDM) and optical time domain multiplexing (OTDM) transmission techniques [1]. Short pulse ultrafast lasers are also necessary for the performance of ultra - high speed, all - optical logic experiments and the demonstration of all – optical Boolean functions such as AND and XOR [2], [3]. Active mode - locking is one of the main methods for the generation of ultra - short transform - limited optical pulses which is based on the direct modulation of the optical field during each cavity round - trip. At 1.5 μm, several actively mode - locked fiber lasers employing Er doped fiber as the gain medium and producing transform - limited picosecond pulses at multi - GHz rates have been demonstrated [4 - 9]. The majority of these systems uses loss modulation by lithium niobate electro - optic modulators due to their large electro - optic coefficient and their compact construction on low loss titanium - indiffused waveguides. Unfortunately, these modulators are highly sensitive to the polarization state of the input optical field. For this reason, laser sources using lithium niobate modulators have to be build from polarization preserving fiber pigtailed components [6], [7] or else suffer from mode - locking loss which limits their operational usability. Similarly, the use of lightly or moderately doped Er fiber results in long cavities which make fiber lasers sensitive to small environmental perturbations, such as thermal fluctuations and acoustic vibration. Although several active stabilization techniques have been developed to counter the instability of long cavity fiber lasers, these increase the complexity of the system [8], [9]. In order to solve these problems, semiconductor optical amplifiers (SOAs) have been deployed in mode - locked laser sources for the generation of short optical pulses at high repetition rates. The use of an SOA inside the laser cavity is particularly attractive since it can provide simultaneously gain over a broad spectral range and modulation due to its fast gain dynamics. Actively mode - locked laser sources with an intracavity SOA have been demonstrated by several research groups [10 - 15] to generate short optical pulses at various repetition rates. In these configurations the SOA provides either the necessary gain exceeding the ring laser cavity loss in combination with an intracavity electroabsorption [10] or electrooptic modulator [11] or both gain and electrically controlled gain modulation [12 - 15].

This paper presents theoretical results obtained from an actively mode - locked fiber ring laser source that uses a single SOA to provide both gain and gain modulation via cross gain saturation from an external optical pulse train. The novel configuration of the fiber ring laser source results in a stable oscillator since the use of a single active element to provide both gain and modulation renders the cavity short. Furthermore, the cavity is nearly polarization insensitive as there is no lithium niobate modulator and the SOA is operated under heavy saturation providing nearly polarization independent gain. The source is very simple to build from commercially available components and has formed a technology platform from which four subsystems of exceptional good operating performance have been demonstrated. These subsystems are crucial for the implementation of high speed telecommunication transmission and network systems that are based on WDM and OTDM techniques. More specifically, these subsystems include (a) a 4.3 ps stable 20 GHz fiber ring laser source with 16 nm tuning range [16], (b) a 10x30 GHz source which generates 10 synchronized wavelength channels, each mode-locked at 30 GHz, producing nearly transform limited 7 ps pulses with less than 5% power variation across them [17], (c) an optical clock repetition rate multiplier circuit which has achieved a factor of 6 frequency multiplication and up to 34.68 GHz [18] and (d) an optical clock recovery and division circuit capable to recover clock from a data pattern up to 20 Gb/s and to perform clock division from the 20 Gb/s stream to 10 and 5 GHz [19]. The four subsystems allow the efficient solution of the following problems that are encountered in WDM and OTDM networks: (a) The generation of ultrashort pulses from a stable tunable source at 40 GHz (fiber ring laser source) to provide the necessary optical power in the ultrafast WDM and OTDM networks, (b) The existence of simultaneous, synchronized wavelength channels (10x30 GHz source). These may be used for clock recovery or synchronization purposes in the complex all optical circuits incorporating digital gates. They can also be deployed in networks combining WDM and OTDM as a single, cost effective source, (c) The synchronization of the all optical gates (AND, XOR) that are used in ultrafast OTDM networks with a single reference optical clock (clock repetition rate multiplier). These gates are key network elements because they perform all-optical functions such as demultiplexing (AND) and pattern matching (XOR) in the users access nodes, (d) The clock recovery (clock recovery/division circuit) operation that is essential for the generation of a high quality local clock synchronized with an incoming data stream to power the optical logic gates and to ensure data synchronization, despite the presence of jittered signal, in very high capacity WDM and OTDM networks.

A mathematical model of the fiber ring laser has been developed describing the mode - locking process in the laser oscillator and providing solutions for the steady - state mode - locked pulse profile. A complete investigation on the critical parameters that determine the width, energy and position of the mode - locked pulse relative to the external pulse has been performed. These parameters include the small signal gain and carrier lifetime of the SOA, the cavity loss and the pulse energy and width of the external modulating signal. From the analysis carried out it has been found that the minimum pulse width is obtained when these parameters are adjusted so that the mode - locked pulse forms in the middle between two successive external pulses and that this condition corresponds to the case where the energies of the external and mode - locked pulses are nearly equal. A rule of thumb has been also derived concerning the relationship between the operating repetition period and the carrier lifetime of the SOA which is particularly important when the fiber ring laser source is used for ultra - high speed all - optical applications.

2. MODEL FOR MODE-LOCKING USING OPTICAL GAIN MODULATION OF A SEMICONDUCTOR OPTICAL AMPLIFIER

Mode - locking is a well known technique for the generation of ultra - short pulses and has been studied theoretically by several research groups [20 - 22]. In particular, active mode - locking has received considerable theoretical interest including analytical [23], [24] and numerical treatment [25], [26] for the estimation of mode - locked pulse characteristics.

This section presents the numerical model that describes the mode - locking process of the SOA fiber ring cavity. The model follows the unidirectional propagation of the pulse around the cavity as it is shown in the block diagram of Figure 1. The operations considered inside the ring cavity are (a) saturable gain in the semiconductor laser optical amplifier (SOA) by the externally introduced pulse and the circulating mode-locked pulse, (b) bandwidth limitation of the circulating pulse with an optical filter and (c) linear losses that are present in the cavity.

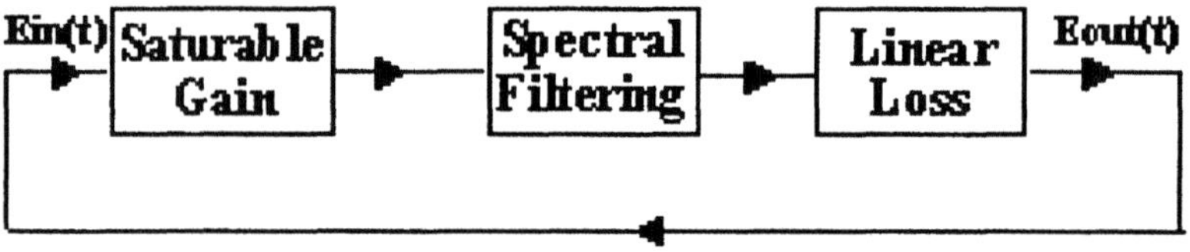

Figure 1. [Block diagram of fiber ring laser source used in simulation]

The parameters of the model are assumed to take values for which the oscillator operates under steady - state mode - locked condition. The pulse is therefore assumed to reproduce itself after one complete circulation around the laser cavity without any changes in the width or the energy, apart from a small temporal displacement. The other implicit assumptions which have been made in order to obtain the solutions are as follows: (a) The electric field of the circulating mode - locked pulse is assumed to be linearly polarized along the SOA axis of peak gain, (b) Changes in the mode - locked pulse profile due to its transit through each cavity element are small. As such, Taylor expansions of exponential terms to the first order incur small errors in the analysis, (c) The pulse energy of the external optical pulse and the circulating mode - locked pulse is assumed to be low enough compared to the saturation energy of the SOA. As a result of this, the gain coefficient of the SOA may be expanded up to second order with respect to pulse energies, (d) The bandwidth of the mode - locked pulse is assumed to be small compared to the bandwidth of the spectral limiting element in the cavity and Taylor expansion to second order of the pulse spectrum in the frequency domain is appropriate.

2.1 Steady – state equations

The profile of the output pulse after the consecutive transit through the SOA, spectral filtering and after encountering the cavity linear losses can be expressed as the successive operation of three operators $\hat{O}_1$, $\hat{O}_2$ and $\hat{O}_3$ on the input pulse profile $E_{in}(t)$

$$E_{out}(t) = \hat{O}_1\hat{O}_2\hat{O}_3\left[E_{in}(t)\right] = \exp\left[-\frac{L}{2}\right]\exp\left[\left(\frac{1}{2\Delta\omega_c^2}\frac{d^2}{dt^2}\right)\right] \cdot \exp\left[\frac{G(t)}{2}\right]E_{in}(t - T_1 - T_2 - T_3) \qquad (1)$$

where T_1 is the transit time of the pulse through the SOA, T_2 is the temporal delay that the pulse experiences when travelling from the exit of the SOA to the exit of the filter and T_3 is the temporal delay that the pulse experiences to travel from the exit of the filter to the entrance of the SOA. In this expression, $\hat{O}_1$ is defined as the gain operator of the SOA acting on the electric field of the input pulses, $\hat{O}_1 \equiv \exp\left[\frac{G(t)}{2}\right]$, with G(t) being the time - dependent power gain coefficient of the SOA, $\hat{O}_2$ is defined as the time domain transfer operator of the pulse through the filter, $\hat{O}_2 \equiv \exp\left[\frac{1}{2\Delta\omega_c^2}\frac{d^2}{dt^2}\right]$, where $\Delta\omega_c$ is related to the full width at half maximum, $\Delta\omega_f$, of $\Delta\omega_c = \frac{1}{2\sqrt{\ln 2}}\Delta\omega_f$ the filter transfer function by and $\hat{O}_3$ is defined as the linear loss operator, $\hat{O}_3 \equiv \exp\left[-\frac{L}{2}\right]$, where L is the power linear loss coefficient of the system.

On the assumption that the profile change per pulse transit is small, the exponential terms in equation (1) may be expanded to first order providing

$$E_{out}(t) = \left[1 - \frac{L}{2} + \left(\frac{1}{2\Delta\omega_c^2}\frac{d^2}{dt^2} + \frac{G(t)}{2}\right)\right] E_{in}(t - T_R) \qquad (2)$$

where $T_R = T_1 + T_2 + T_3$ is the round trip time of the laser cavity, which for simplicity is assumed to be equal to the repetition period, T_{ext}, of the externally introduced pulse train.

In order to obtain closure and a steady - state solution for the mode – locked pulse, the mode - locked pulse is assumed to reproduce itself after each complete transit through the SOA, filter and losses. The gain of the SOA is also assumed to recover to the same value before each mode - locked pulse enters it. In order to accommodate small temporal shifts, δT, of the pulse around its peak as a result of the gain saturation in the SOA, the self - reproducing pulse profile principle is applied by requiring that $E_{out}(t) = E_{in}(t - T_R + \delta T)$ and by expanding to first order in δT we obtain from (2)

$$\left(1 - g(t) - \frac{1}{L}\frac{1}{\Delta\omega_c^2}\frac{d^2}{dt^2} + \frac{\delta}{\Delta\omega_c}\frac{d}{dt}\right) E_{in}(t) = 0 \qquad (3)$$

where we have introduced the normalized parameters $\delta = \frac{2}{L}\Delta\omega_c \delta T$ and $g(t) = \frac{G(t)}{L}$

Specifically, g(t) is the gain coefficient of the SOA normalized to the linear loss coefficient and δ is a time delay (or advance) parameter indicating whether the pulse repetition period deviates from the free space round trip time T_R.

The solution of this steady - state equation for the recirculating mode - locked pulse requires the calculation of the time dependent normalized gain coefficient g(t). For this purpose, we consider in 2.2 the gain dynamics of the SOA during its gain saturation and recovery periods. In the analysis that follows, the external pulses are temporally separated from the mode - locked pulses by an unknown parameter, T_s, that is assumed to be large compared to the width of the two pulses. Physically, this means that the external pulse energy does not play any direct role in saturating the gain of the SOA during the presence of the mode - locked pulse and therefore need not be taken into account in equation (3) explicitly. The effect of the saturation of the SOA by the external pulses is however taken into account in calculating the input gain parameter at the entrance point of the mode - locked pulses as will be described in 2.4.

2.2 Semiconductor optical amplifier gain dynamics

The ordinary differential equation that governs the temporal saturation for each point in the pulse is [27]

$$\frac{dg(t)}{dt} = \frac{g_{ss} - g(t)}{t_{car}} - \frac{cn\varepsilon_o A}{2} \frac{|E_{in}(t)|^2}{J_{sat}} \{\exp[g(t)] - 1\} \quad (4)$$

where $g_{ss} = \Gamma\sigma N_o(I/I_o - 1)/L$ is the integrated small signal gain normalized to the cavity linear losses and $J_{sat} = \hbar\omega_o A/\sigma$ is the saturation energy of the amplifier. The parameters t_{car}, ε_o, A, Γ, σ, $\hbar\omega_o$, N_o and I_o are the carrier lifetime, the dielectric constant in vacuum, the cross section area of the active region, the confinement factor, the transition cross section, the photon energy and the carrier density and injection current required for transparency, respectively.

2.3 Gain saturation by short optical pulse

Equation (4) can be now used to obtain the gain saturation of the SOA for a short optical pulse, on the assumption that the width of the pulse is much smaller than the carrier lifetime of the amplifier [27]. This in practice means that the optical pulse is so short that the gain has no time to recover during its duration and the first term in the right hand side of equation (4) can be thus neglected. Assuming also for simplicity that around the peak of the mode - locked pulse g(t) is close to 1 so that the exponential term may be expanded to first order, the resulting normalized time dependent gain g(t) is

$$g(t) = g_{in}^{P} \exp\left[-\frac{J_{in}(t)}{J_{sat}}\right] \quad (5)$$

where g_{in}^{p} is the gain before the arrival of the recirculating mode – locked optical pulses and $J_{in}(t) = \frac{cn\varepsilon_0 A}{2} \int_{-\infty}^{t} |E_{in}(t')|^2 dt'$ is the instantaneous energy of the mode - locked pulses entering the SOA.

The exponential of (5) may be expanded, according to the assumptions already stated, into second order Taylor series and substituting in (3) yields

$$\left[1 - g_{in}^{p} + g_{in}^{p} \frac{J_{in}(t)}{J_{sat}} - \frac{g_{in}^{p}}{2}\left(\frac{J_{in}(t)}{J_{sat}}\right)^2\right] E_{in}(t) + \frac{\delta}{\Delta\omega_c}\frac{dE_{in}(t)}{dt} - \frac{1}{L}\frac{1}{\Delta\omega_c^2}\frac{d^2 E_{in}(t)}{dt^2} = 0 \quad (6)$$

The solution of this equation requires the knowledge of g_{in}^{p} which can be found from the gain recovery analysis that follows.

2.4 Gain recovery

After the saturation of the SOA according to (5), the gain recovers due to the injection of carriers by the injection current. The gain recovery can be calculated, assuming that the stimulated recombination term (second term) in the right hand side of (4) can be neglected, from

$$g(t) = (g_f - g_{ss}) \exp\left(-\frac{t}{t_{car}}\right) + g_{ss} \quad (7)$$

where g_f is the gain of the saturated SOA immediately after the transit of a pulse through given by

$$g_f = g_{in} \exp\left[-\frac{J_{total}}{J_{sat}}\right] \quad (8)$$

In this equation g_{in} and J_{total} are the SOA gain and pulse energy immediately before the pulse entrance into the SOA. Given that two different sets of pulses transit the SOA at different times, i.e. the external and mode - locked pulses, the recovery of the SOA must be accounted separately for these sets of pulses. For this reason, g_{in} takes the values g_{in}^{p} and g_{in}^{ext} and J_{total} takes the values J_p and J_{ext} for the mode - locked and external pulses, respectively

In order to obtain the steady - state condition for the laser oscillator, the gain of the SOA is assumed to recover always to the same level g_{in}^{p} and g_{in}^{ext} before the mode - locked and the external pulses enter it, respectively. As the mode - locked pulse forms at time T_s after the external pulse has entered the SOA, its gain has had $T_s - \Delta T$ time to recover to the value g_{in}^{p} from the value g_f^{ext} before the arrival of the mode - locked pulse, where $\Delta T = \frac{t_{text} + t_p}{2}$ with t_{ext} being the width of the external pulse.

Similarly, the external pulse experiences a gain equal to g_{in}^{ext} which has recovered from the saturated value g_f^{p} after the mode - locked pulse exits the SOA after $T_R - T_s - \Delta T$ time has elapsed. The equation of the recovery of the SOA is thus split in two equations which describe the SOA recovery after the transition of the external and mode - locked pulses, respectively. Equations (7) and (8) yield

$$g_{in}^{p} = \left[g_{in}^{ext} \exp\left(-\frac{J_{ext}}{J_{sat}} \right) - g_{ss} \right] \exp\left(-\frac{T_s - \Delta T}{t_{car}} \right) + g_{ss} \tag{9}$$

and

$$g_{in}^{ext} = \left[g_{in}^{p} \exp\left(-\frac{J_p}{J_{sat}} \right) - g_{ss} \right] \exp\left(-\frac{T_R - T_s - \Delta T}{t_{car}} \right) + g_{ss} \tag{10}$$

Substituting (10) in (9), g_{in}^{ext} is eliminated and an expression is obtained which involves only g_{in}^{p}

$$g_{in}^{p} = \frac{g_{ss}[1+\exp(A-\alpha)-\exp(A+B-\alpha)-\exp(A)]}{[1-\exp(A+B-U_o-\alpha)]} \quad (11)$$

where $U_o = \frac{J_p}{J_{sat}}$ and $\alpha = \frac{J_{ext}}{J_{sat}}$ are the normalized to the SOA saturation energy energies of the mode - locked and external pulses

respectively, $A = -\frac{T_s - \Delta T}{t_{car}}$ and $B = -\frac{T_R - T_s - \Delta T}{t_{car}}$.

2.5 Solution

In order to obtain a steady - state solution for the mode - locked pulse profile in terms of its width and energy, the mode - locked pulses are assumed to be squared hyperbolic secant in shape with total energy J_p so that the electric field is given by

$$E_{in}(t) = \left(\frac{U_o J_{sat}}{2t_p}\right)^{1/2} \operatorname{sech}\left(\frac{t}{t_p}\right) \quad (12)$$

where t_p is related to the full width at half maximum, T_{FWHM}, by

$$t_p = \frac{T_{FWHM}}{\ln(3+2\sqrt{2})}.$$

Squaring (12) and integrating over the time dependent variable we obtain

$$\frac{J_{in}(t)}{J_{sat}} = \frac{U_o}{2}\left(1+\tanh\frac{t}{t_p}\right) \quad (13)$$

Substituting (12) and (13) and equating the coefficients of the same order terms in (6), the following equations are derived

$$1 - g_{in}^{p} + \frac{g_{in}^{p}\,U_o}{2} - \frac{g_{in}^{p}\,U_o^{2}}{8} - \frac{1}{L}\frac{1}{t_p^2 \Delta\omega_c^2} = 0 \tag{14}$$

$$\frac{g_{in}^{p}\,U_o}{2} - \frac{g_{in}^{p}\,U_o^{2}}{4} - \frac{\delta}{t_p\,\Delta\omega_c} = 0 \tag{15}$$

$$\frac{2}{L t_p^2\,\Delta\omega_c^2} - \frac{g_{in}^{p}\,U_o^{2}}{8} = 0 \tag{16}$$

Equations (11), (14), (15) and (16) form a set of nonlinear equations with unknowns t_p, U_o, g_{in}^{p}, and δ. Setting $\delta = \frac{x}{L t_p \Delta\omega_c}$ we obtain

$$g_{in}^{p} = \frac{(x+4)^2}{(x+4)^2 - 4(x+1)} \tag{17}$$

$$U_o = \frac{8}{x+4} \tag{18}$$

$$t_p \Delta\omega_c = \frac{|x+4|}{2\sqrt{g_{in}^{p} L}} \tag{19}$$

Equations (11), (17), (18) and (19) form the final set of equations for the calculation of $t_p \Delta\omega_c$.

3. RESULTS OF THE SELF-CONSISTENT MODEL AND DISCUSSION

3.1 Calculation of $T_{FWHM}\Delta\omega_f$

The calculation of $t_p\Delta\omega_c$ and is performed by setting equation (11) equal to (17) and substituting the known values of the period and pulse width of the external pulse train, the SOA carrier lifetime and the cavity loss. The derived equation in x is however transcendental and requires numerical solution. In order to solve the transcendental equation and find the unknown x, the parameters α and T_s are scanned in accordance with the experimental conditions for a given value of the normalized small signal gain, g_{ss}, which is the only independent variable of the nonlinear system of equations. The unknown x is also scanned within a range whose lower limit is determined by the fact that the normalized energy of the mode - locked pulses must be always less than the normalized energy of the external pulses. Note that due to the form of the transcendental equation, the same combinations of the values of U_o and α may correspond to multiple solutions of x. The selection of the single x solution and the elimination of the others is performed by choosing the value that corresponds to the smallest pulse width. Therefore, by solving the system for minimum pulse width, the optimum value of T_s and α is obtained for the given g_{ss}. The value of x that corresponds to this set of g_{ss}, α and T_s is then replaced in (19) to calculate $t_p\Delta\omega_c$ and the desired value of $T_{FWHM}\Delta\omega_f$ is finally obtained from

$$\Delta\omega_c = \frac{1}{2\sqrt{\ln 2}}\Delta\omega_f \quad \text{and} \quad t_p = \frac{T_{FWHM}}{\ln(3+2\sqrt{2})}$$

3.2 Simulation results and discussion

Simulation results were obtained for repetition frequencies at 10 GHz and 20 GHz in order to compare them to the experimental ones. The parameters used for the simulations were chosen to agree with the values in the experimental configuration [16]. Specifically, the width of the external pulses was 8 ps, the cavity loss was 15 dB and the full width at half maximum of the spectral limiting filter was 5 nm for both frequencies. The small signal power gain coefficient that was used in the calculation of the normalized to the loss small signal gain, g_{ss} could take values in the range 19.2 - 25.2 dB, in accordance to the gain characteristic of the SOA. The values also of the carrier lifetime were selected to be 100 ps at 10 GHz and 70 ps at 20 GHz in order to match the theoretical and experimental data provided for the 500 μm SOA that was used in the experiment. The decrease of the carrier lifetime at the higher repetition frequency is a result of the higher current at which the SOA has to be driven [28], [29] to account for twice the number of pulses entering the amplifier. The carrier lifetime is a particularly important parameter because it defines the range of repetition frequencies at which the ring laser may be operated for a given SOA, as it will be discussed with the simulation results at 20 GHz.

Figure 2 (a) shows the variation of the pulse width, T_{FWHM}, and the normalized energy, U_o, of the mode - locked pulse as well as of the required external pulse energy, α, against the normalized small signal gain coefficient, g_{ss}, at 10 GHz. This figure shows that as the normalized small signal gain increases, the energy of the mode - locked pulse also increases saturating the SOA more strongly and leading to pulse narrowing. Similarly, the energy of the external pulse must also be increased to preserve the condition of shortest pulse from the oscillator. The width of the shortest mode - locked pulse decreases up to a minimum value as the normalized gain increases. The minimum pulse width was obtained for 1.53 normalized small signal gain and was 4.5 ps assuming a filter bandwidth of 5 nm in the experimental system. The value of the mode - locked pulse energy, U_o, is either slightly less than or nearly equal to that of the external pulse, α. At the point of minimum pulse width, U_o takes the value 0.45 and nearly equals α which is 0.46. The SOA is already heavily saturated at the same point and a further increase in the gain beyond this point requires a further increase in the energy of the external pulse, which does not allow an increase in the energy of the mode - locked pulse and results in an increase of its pulse width.

Figure 2 (b) illustrates the variation of the temporal separation, T_s, between the mode - locked and external pulses and for convenience the pulse width, T_{FWHM}, of the mode - locked pulses versus the normalized small signal gain, g_{ss}, at 10 GHz. This figure shows that provided the conditions for best pulse width are maintained, the mode - locked pulse forms at 48 ps, approximately half way between two successive external pulses, despite large variations in g_{ss}. As the small signal gain increases, the mode - locked pulse moves away from the half way point and closer to the leading of the external pulses. This is due to the fact that as the small signal gain increases, the gain recovers above the losses in the cavity earlier, forcing the mode - locked pulse to move forward. However, the excess gain increases very gradually so that the mode - locked pulse moves very slowly towards the leading of the two consecutive external pulses.

From the observation and analysis of Figures 2 (a) and (b) it can be deduced that two conditions must be fulfilled in order to obtain mode - locked pulses with the minimum possible width. The first is that the normalized energies of the mode - locked and external pulses must be nearly equal. This condition alone is not sufficient since there is a large set of parameter values for which the normalized energies are equal but the pulse width is not globally minimum as seen in Figure 2 (b). This leads to the second condition which states that the mode - locked pulse must form nearly equidistantly between consecutive external pulses. When both conditions hold, the fiber ring laser system generates the shortest pulses. Figure 2 (c) displays the variation of the time delay, T_s, between the mode - locked and external pulses against the energy of the external pulse normalized to the saturation energy of the SOA. The results shown are for operation of the laser at 10 GHz with the normalized small signal gain set at 1.53. The figure shows that as the energy of the external pulse increases, the mode - locked pulse becomes increasingly delayed with respect to the leading of the two successive external pulses. This is a consequence of the deeper saturation into which the SOA is brought by the external pulse as its energy increases, which in turn requires a longer recovery time for the gain of the SOA to overcome the cavity losses and allow the mode - locked pulse to form. In the same figure the variation of the pulse width against the normalized external pulse energy has been plotted. As seen, the pulse width reaches a minimum value for $\alpha = 0.46$ and forms roughly half way between two consecutive external pulses. Further increase of the external pulse energy beyond this point delays the formation of the mode - locked pulse further and results in its broadening.

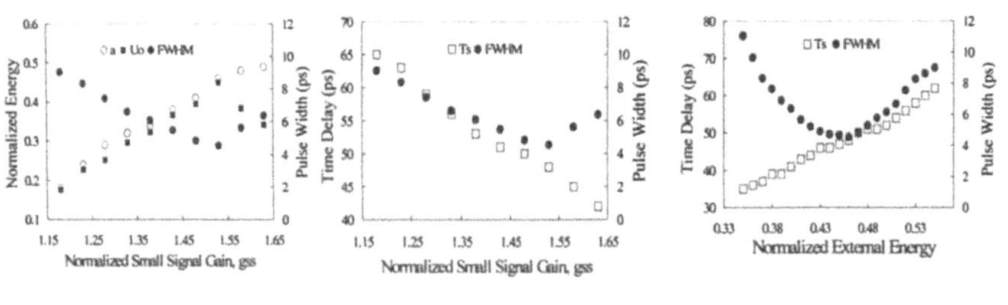

Figure 2. 10 GHz simulation. (a) Variation of the mode – locked pulse width, FWHM, the normalized energy of the mode – locked , U_o, and external pulses, T_s, versus the normalised small signal gain, g_{ss} . (b) Variation of the time separation between the mode – locked and external pulses, T_s ,versus the normalized small signal gain, g_{ss}. (c) Variation of the time separation between the mode – locked and external pulses, T_s, versus the normalized energy of the external pulses, α.

The model was next used to evaluate the performance of the laser system as the repetition frequency of the external pulse train was increased to 20 GHz. In order to examine the behaviour of the model, initially the same values for the physical parameters were used as in the case for 10 GHz operation. This time, however, the width of the mode - locked pulses was longer than that of the external pulses. Note that an increase in the repetition rate of the oscillator restricts the range of values that T_s may take, so that there is no overlap between the external and mode - locked pulses. The parameter that most critically determines the quality of mode - locking, however, is the carrier lifetime of the SOA. If the recovery time is far too short for a given repetition period, the gain of the SOA recovers too fast so that the mode - locked pulse forms early behind the external pulse. This allows excess gain to build at its trailing edge, resulting in longer mode - locked pulses. On the contrary, if the recovery time is far too long, the mode - locked pulse trails too far behind the external pulse and has insufficient energy to strongly modulate the gain of the SOA, being therefore long again. In order to obtain a rule of thumb for the carrier lifetime, the theory of standard mode - locking is considered which stipulates that mode - locked pulses form in a laser cavity if the normalized gain function g(t) is greater than 1 just as the pulse forms. In all other instances, that is ahead and behind the pulse, g(t) must be smaller than 1. In this sense and so as to ensure pulse formation, the carrier lifetime must be selected so that $g_{in}^{p} > 1$ and $g_{f}^{p} < 1$. Applying first and second order Taylor expansion approximations in (5) and (11), these inequalities hold simultaneously if

$$1.1(T_R - 2t_{ext}) < t_{car} < 2.1(T_R - 2t_{ext}) \tag{20}$$

which sets a lower and an upper limit for the amplifier carrier lifetime with respect to the repetition period and the width of the external pulse. This condition implies that for the mode - locked pulses to form, the carrier lifetime of the SOA must decrease as the operating frequency of the ring laser increases. The inverse relationship between the carrier lifetime and the operating frequency may be explained if the roles of the frequency and small signal gain in the mode - locking process are considered. Specifically, an increase of the operating frequency requires an increase of the small signal gain in order to compensate the decrease in energy of the mode - locked pulses due to the larger number of pulses entering the amplifier per unit time. Experimentally, an increase of the small signal gain of the SOA is achieved with a higher drive current which also results in a decrease of its carrier lifetime [28], [29]. Following the preceding analysis, the carrier lifetime used in the 20 GHz simulation was set to 70 ps.

Figure 3 (a) depicts the variation of the pulse width and the normalized energy of the mode - locked and external pulses as the normalized small signal gain coefficient is varied at 20 GHz. The behaviour of the oscillator is very similar to that presented in Figure 2 (a) for the 10 GHz case. The minimum pulse width obtained was 3.9 ps for an increased normalized small signal gain of 1.83. Despite the increase in the small signal gain, the energies of the mode - locked and external pulses were lower than the 10 GHz case, due to the heavy saturation of the SOA, but were again nearly equal at the point of minimum pulse width ($U_o = 0.36$, $a = 0.37$). Note that if the same value of small signal gain as for the 10 GHz operation is used, the mode - locked pulses are longer, 7 ps, with lower energy, $U_o = 0.22$.

Figure 3 (b) shows the variation of the time separation between the mode - locked and external pulses, T_s, versus g_{ss}. The pulse width of the mode - locked pulses is also included so as to compare with the results from Figure 2 (b). The variation of the position of the mode - locked pulse at 20 GHz has similar behaviour as at 10 GHz, that is, as the small signal gain increases, the mode - locked pulse moves forward towards the leading external pulse. The minimum pulse width is obtained for $T_s = 25$ ps, half way between two consecutive external pulses.

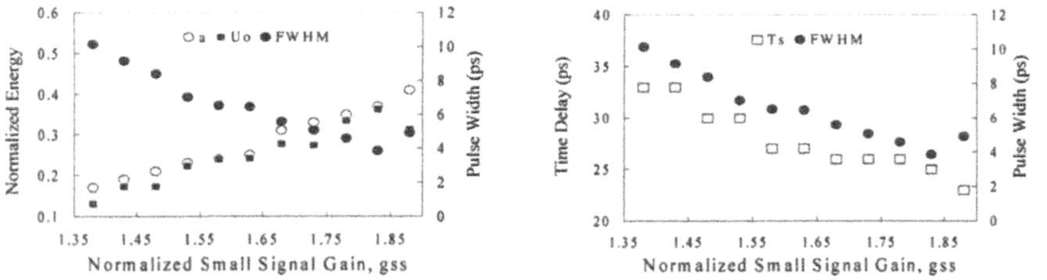

Figure 3. 20GHz simulation. (a) Variation of the mode – locked pulse width, FWHM, the normalized energy of the mode – locked, U_0, and external pulses, α, versus the normalized small signal gain, g_{SS}. (b) Variation of the time separation between the mode – locked and external pulses, T_s, versus the normalized small signal gain, g_{SS}.

In order to assess the ability of the model for quantitative prediction, the parameter values from the experimental set up were substituted in equations (11), (14), (15) and (16). At 10 GHz, with an experimental value for g_{ss} 1.53 and using a filter of 5 nm spectral bandwidth and external modulating pulses of 8 ps width, the model predicts mode - locked pulses of 4.5 ps duration with normalized energy 0.45 and external pulses with normalized energy 0.46. In this instance the mode - locked pulse forms 48 ps behind the external pulse. At 20 GHz, for an increased normalized gain of 1.78, the model predicts mode - locked pulses of 4.5 ps duration and 0.34 normalized energy forming 26 ps behind the external pulses of 0.35 normalized energy. Experimentally, the shortest pulse train had 4.3 ps duration and formed 49 ps behind the external pulses at 10 GHz. At 20 GHz, the corresponding values were again 4.3 ps and 24 ps delay behind the external pulses. The agreement thus between the experimental and the theoretical values for the pulse width and the relative delay is exceptionally good given the simplicity of the model and suggests its robustness to the approximations that necessarily had to be made.

4. CONCLUSION

We have presented a theoretical study of the performance of a high repetition rate laser source that exploits the fast saturation of an intracavity semiconductor optical amplifier, forming the gain element in its cavity, from an external optical pulse to generate ultrashort pulses. A complete and computationally simple numerical model has been developed to provide solutions for the steady - state mode - locked pulse profile in terms of the critical parameters of the system. The results indicate that the minimum pulse width is obtained when these parameters are adjusted so that the mode - locked pulse forms in the middle between two successive external pulses and that this condition corresponds to the case where the energies of the external and mode - locked pulses are nearly equal. A practical rule of thumb has been also derived concerning the optimum combination of the SOA carrier lifetime and the repetition frequency that is required for the novel ring laser to be used as a high repetition rate source for transmission or optical logic experiments. The comparison of the experimental and theoretical results has revealed an excellent agreement between them and has helped to provide a thorough explanation of the mode - locking process.

5. REFERENCES

[1] S. Kawanishi, H. Takara, K. Uchiyama, I. Shake, K. Mori, Electron. Lett. 35 (1999) 826.
[2] K.L. Hall, K.A. Rauschenbach, Opt. Lett. 23 (1998) 1271.
[3] K. Zoiros, T. Houbavlis, K. Vlachos, H. Avramopoulos, F. Girardin, G. Guekos, S. Hansmann, H. Burkhard, CLEO Techn. Dig. CThF5 (1999) 379.
[4] J. D. Kafka, T. Bauer, D.W. Hall, Opt. Lett.14 (1989) 1269.
[5] D. Jones, H. Haus, E. Ippen, Opt. Lett. 21 (1996) 1818.

[6] T. Pfeiffer, G. Veith, Electron. Lett. 29 (1993) 1849.
[7] E. Yoshida, Y. Kimura, M. Nakazawa, Electron. Lett. 31 (1995) 377.
[8] G.T. Harvey, L.F. Mollenauer, Opt. Lett. 18 (1993) 107.
[9] X. Shan, D.M. Spirit, Electron. Lett. 29 (1993) 979.
[10] M.J. Guy, J.R. Taylor, D.G. Moodie, A.E. Kelly, Electron. Lett. 32 (1996) 2240.
[11] T. Papakyriakopoulos, A. Stavdas, E.M. Protonotarios, H. Avramopoulos, Electron. Lett. 35 (1999) 717.
[12] S.W. Corzine, J.E. Bowers, G. Przybylek, U. Koren, B.I. Miller, C.E. Soccolich, Appl. Phys. Lett. 52 (1988) 348.
[13] R.S. Tucker, U. Koren, G. Raybon, C.A. Burrus, B.I. Miller, T.L. Koch, G. Eisenstein, Electron. Lett. 25 (1989) 621.
[14] H. Shi, G.A. Alphonse, J.C. Conolly, P.J. Delfyett, Electron. Lett. 34 (1998) 179.
[15] C. Schimdt, E. Dietrich, S. Diez, H.J. Ehrke, U. Feiste, L. Kuller, R. Ludwig, H.G. Weber, CLEO Techn. Dig. CThA3 (1999) 348.
[16] T. Papakyriakopoulos, K. Vlachos, A. Hatziefremidis, H. Avramopoulos, Opt. Lett. 24 (1999) 1209.
[17] K. Vlachos, K. Zoiros, T. Houbavlis, H. Avramopoulos, to appear in IEEE Photon. Technol. Lett., Jan. 2000
[18] T. Papakyriakopoulos, K. Vlachos, A. Hatziefremidis, H. Avramopoulos, Opt. Lett. 24 (1999) 717.
[19] K. Vlachos, T. Papakyriakopoulos, T. Houbavlis, H. Avramopoulos, ECOC Proc. (1999) I – 414.
[20] D.J. Kuizenga, A.E. Siegman, IEEE J. Quantum Electron. 6 (1970) 694.
[21] G.H.C. New, IEEE J. Quantum Electron. 10 (1974) 115.
[22] H.A. Haus, Japan. J. Appl. Phys. 20 (1981) 1007.
[23] H.A. Haus, J. Appl. Phys. 51 (1980) 4042.
[24] M. Schell, A.G. Weber, E.H. Boettcher, E. Scholl, D. Bimberg, IEEE J. Quantum Electron. 27 (1991) 402.
[25] M. Schell. A.G. Weber, E. Scholl, D. Bimberg, IEEE J. Quantum Electron. 27 (1991) 1661.
[26] P.J. Delfyett, A. Dienes, J.P. Heritage, M.Y. Hong, Y.H. Chang, Appl. Phys. B 58 (1994) 183.
[27] G. Agrawal and N.A. Olsson, IEEE J. Quantum Electron. 25 (1989) 2297.
[28] F. Girardin, G. Guekos, A. Houbavlis, IEEE Photon. Technol. Lett. 10 (1998) 784.
[29] G. Eisenstein, R.S. Tucker, J.M. Wiesenfeld, P.B. Hansen, G. Raybon, B.C. Johnson, T.J. Bridges, F.G. Storz, C.A Burrus, Appl. Phys. Lett. 54 (1989) 454.

Reduction of Hop-Count in Packet-Switched Networks using Wavelength Reconfiguration

S. N. Larsson*, S. Hubendick.
*Telia Research AB, S-123 86 Farsta, Sweden (stefan.n.larsson@telia.se), *also with Royal Institute of Technology.*

Keywords: Virtual Topology, Optical Cross-Connect, Network Reconfiguration, WDM.

Abstract: A method using packet flows to create a wavelength topology is shown to reduce the hop-count in a packet-switched network. The wavelength cross-connect function is achieved by opaque cross-connects, providing full wavelength conversion. Simulations on a 5-node, an 11-node and a 26-node network are done for different number of ports per node and wavelength channels per link.

1. INTRODUCTION

The use of optical layer functionality is still in its infancy. DWDM has since the mid-nineties been widely deployed by Telcos to overcome fibre exhaust. Apart from using WDM as a capacity booster, other optical layer functions are uncommon. The fact that optical layer functionality on the wavelength level is very simple and potentially cheap has however lead to increasing interest in optical network elements such as optical add/drop multiplexers (OADM) and optical cross-connects (OXC). While static OADMs are becoming deployed as a means of reducing network cost, /1/ the reason for using OXCs is presently mainly due to protection /2/ or reduction of personnel /3/. However, new optical layer functions using OXCs for traffic engineering (load balancing and redirection of traffic) and Class of Service are proposed /4,8/.

In order to implement the OXC a technology choice must be made. First, should the OXC work entirely in the optical domain, i.e. optically transparent, or is optical/electrical conversion allowed? If a transparent OXC is desired, common switch-technologies include opto-mechanical, thermo-optical (e.g. polymers), electro-optical (e.g. $LiNbO_3$) or acousto-optical. Of these, mainly the opto-mechanical has gained commercial interest, while new micro-mechanical switches (MEMS) are emerging as a strong candidate /5/. Transparent OXCs always impose some limitation on the optical network, e.g. maximum size or capacity. Opaque OXCs (including o/e conversion) can to a large extent relax size/capacity issues by using standard electrical regeneration. To assure flexible networking, the regeneration should be multi-rate and such circuits are commercially available today. Apart from being small in size, electrical switch-cores are cheap, can be made with a large number of ports and exhibit fast switching speeds (10's of ns). The drawback with electrical switch-cores is the limit on maximum bit-rate per port. 2.5 Gb/s is the limit of commercial products today.

In this paper, a method for using opaque OXCs to connect packet-switches in a dynamic way is proposed. The logical network topology[1] is adapted to the traffic demand by an optical network management system. The key questions in this study are:

1. how much can one benefit from having a re-configurable physical network when traffic demands change[2]
2. how this benefit depends on number of nodes, router ports and wavelengths per fibre.

The gain is measured as the reduction in number of router hops. A field trial evaluating the OXC implementation and optical layer functionality is described elsewhere /6/.

2. NODE AND NETWORK MODEL

Figure 1 shows the node model used. On top of the (opaque) OXC is the packet switch (e.g. IP-router) interfacing the OXC with a number of optical ports. The OXC is characterised by the number of input (=output) ports which is equal to the number of wavelength channels of all input links plus the number of add (=drop) switch-ports. With this OXC, wavelength conversion is performed in every node. In the following, the number of ports refers to the switch-ports and not all of the ports of the OXC.

[1] The virtual topology consisting of reconfigurable wavelength channels connecting the switches (or routers) via the OXCs.

[2] A similar study is presented in /9/ for a 14-node network.

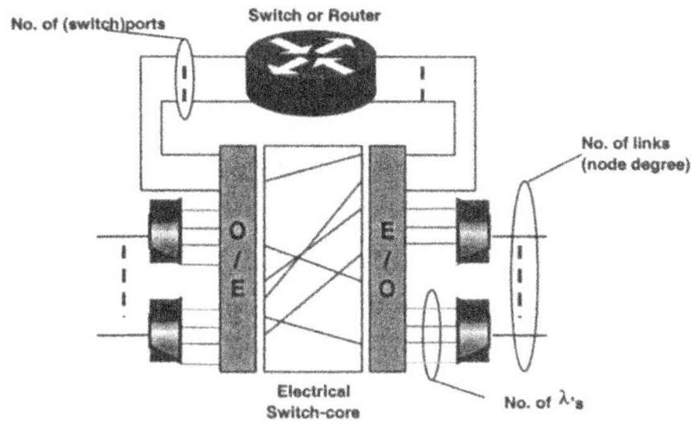

Figure 1 The node model

The physical network models are shown in figure 2. with b) originating from the COST239 project /7/.

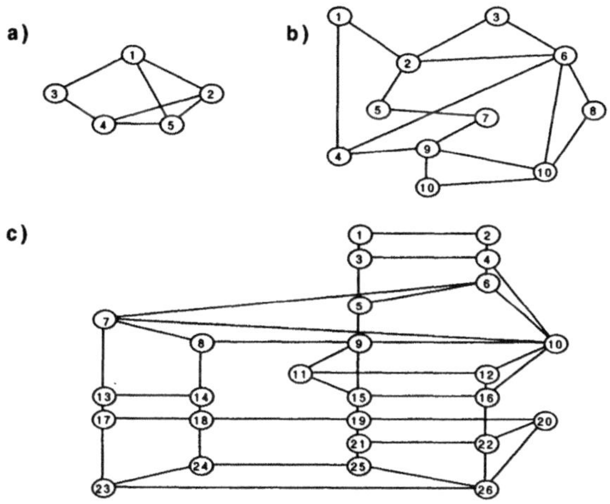

Figure 2 The physical network models. a) 5-node, b) 11-node and c) 26-node. Each node includes an OXC.

The average node degrees of the three networks a-c) are 2.80, 2.91 and 3.46 respectively. Thus, all networks used are meshed to varying degrees.

3. EXAMPLE

The simulated traffic demands are randomly generated as a uniformly distributed variable from 0 to 100 uni-directionally for each node pair. Table

1 shows an example for the 5-node network. (The demand from node i to node i is set to 0)

To\From	Node 1	Node2	Node3	Node 4	Node 5
Node 1	0	99	90	11	78
Node 2	88	0	8	92	57
Node 3	16	35	0	14	67
Node 4	45	21	2	0	88
Node 5	93	32	100	17	0

Table 1 Traffic demands by randomly generated number of packets.

The demands are then summed for each pair and ordered so that a list of demands is given. (E.g. the combined demand from node 4 to node 5 is 88+17=105.) The resulting list would be:

Demand #	Node nr.	Node nr.	# of packets
1	1	2	187
2	1	5	171
3	3	5	167
4	2	4	113
5	1	3	106
6	4	5	105
7	2	5	89
8	1	4	56
9	2	3	43
10	3	4	16

Table 2 Ordered list of traffic demands from table 1.

From this list the bi-directional wavelength connections in the network are made in sequence using the Dijkstra algorithm for shortest path calculation as described by the psuedo-code in figure 3.

```
OptimizeNet_Simple(physical_Net,Demands)
logical_NET ← null;
for k ← 1 to NumberOf(Demands)
 do  node_a,node_b ← Demands(k)
      if connected(NET,node_a,node_b) & exist_port(node_a)& exist_port(node_b)
       then sp ← connect_shortest_path(NET,node_a,node_b)
            add_path(logical_NET,sp)
check_full_connectivity(logical_NET)
return logical_NET
```

Figure 3 The pseudo-code of the proposed method.

Figure 4 show the physical and logical networks and the paths through the network when using table 2. Here, the number of ports per node is set to two. As a result demand 5 is excluded since node 1 used its ports for demands 1 and 2. However, bypass traffic through node 1 can still be handled if free WDM-channels are available. Demands 6-9 are also excluded due to the same reason. Thus five demands are satisfied, which equals half the number of nodes times the number of ports per node. This characterises the proposed method: the demands with largest traffic are assigned a wavelength channel until either the ports or WDM channels are depleted and the other demands must go through one or more additional switches (routers) to reach their destination.

The average number of hops is now given as the number of packets in table 2 times their respective number of hops divided by the total number of packets. Demands 1-4 and 10 are single-hop while the rest are two-hops. This gives: ((187+171+167+113+16)+(106+105+89+56+43)*2)/1053= <u>1.3789</u>

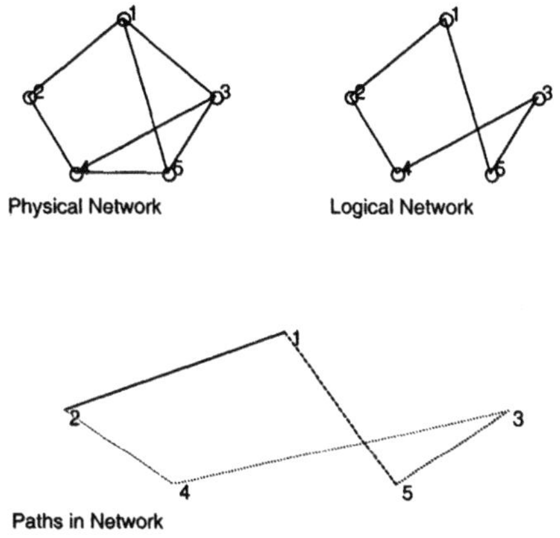

Figure 4 The five-node physical network and the logical network and paths resulting from table2.

A new random demand is shown in table 3.

To\From	Node 1	Node2	Node3	Node 4	Node 5
Node 1	0	63	93	8	15
Node 2	29	0	71	6	24
Node 3	77	32	0	6	24
Node 4	15	35	16	0	87
Node 5	62	26	71	24	0

Table 3 New randomly generated number of packets.

If now the average number of hops is calculated using the logical network of figure 4, the result for table 3 would be <u>1.5106</u> which is an increase of 9.6 % compared to the result when optimising the logical network from table 2.

Some list of demands result in logical networks that are not connected. E.g. if the demand list for the 5-node network, with 2 ports per node, started with 1-2, 1-4 and 2-4, the nodes 3 and 5 can only communicate with each other and the network is not connected. The algorithm in Figure 3 checks for such cases and if so, a re-start is made.

4. RESULTS

In all cases the number of ports per node is lower then the (number of nodes-1) in the network. Thus, full wavelength-connectivity in the networks can not be achieved and benefit from re-configuring the wavelength connections is possible (more than one wavelength channel per node pair is not considered). In the following, wavelength connections are referred to as "logical connections".

Starting with the 5-node network, average hop-count for 4 random demands was calculated resulting in 4 optimised logical networks. (The reason for choosing 4 demands is a compromise between readable graphs and the need for sufficient statistics). On these 4 optimised logical networks, average hop-count for the other demands (i.e. not the ones used for optimising the logical network) was calculated resulting in 4*(4-1)=12 values. The top graph of figure 5 show these results for 2 and 3 ports/node with 3 WDM-channels per link. Each star represents an optimised case and the solid line connects the two numbers of ports for that demand. The circles and dashed lines represent the non-optimised results. The bottom graph of figure 5 show the non-optimised results in percentage increase relative to the optimised case.

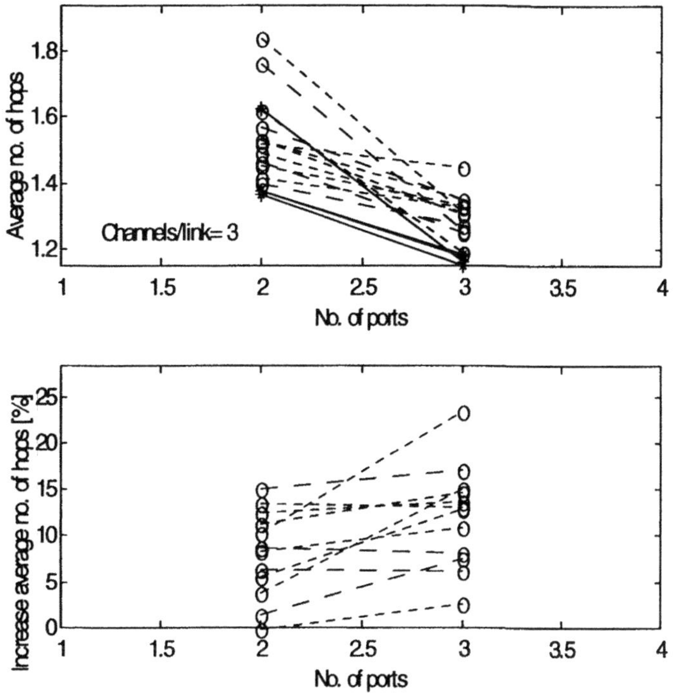

Figure 5 Average number of hops (top) and increase in average number of hops relative to the optimised case (bottom) for 4 demands. Stars are for optimised results and circles for non-optimised.

Another way of representing the result for optimised and non-optimised logical networks for different demand is by using histogram plots. Figure 6 show the number of packets vs. the number of hops for 4 demands. The left graph show the result for an optimised logical network (using demand 1) and the three others are for different demands applied to the logical network optimised according to demand 1.

The same calculations are made for the 11-node network with the results showing in figures 7,8 for 4 WDM-channels per link, figures 9,10 for 6 channels per link and figures 11,12 for 8 channels per link. To assure connected logical networks the minimum number of ports is 3.

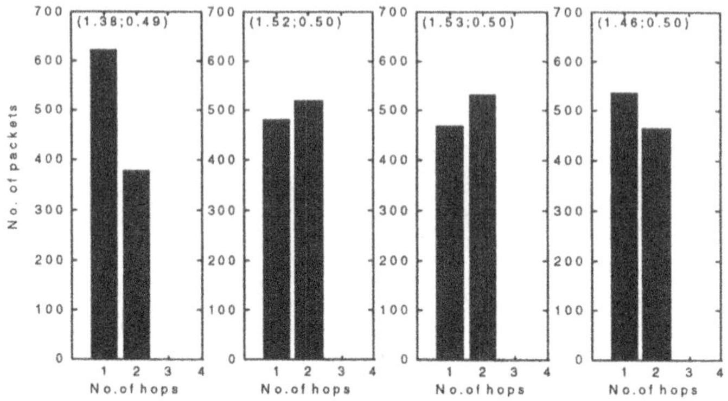

Figure 6 Number of hops vs number of packets for an optimised logical network (left) using one demand and for 3 other demands applied to that logical network (i.e. non-optimised). Ports=2 and WDM channels/link=3. The average and standard deviation are shown in each graph as (Av;Std).

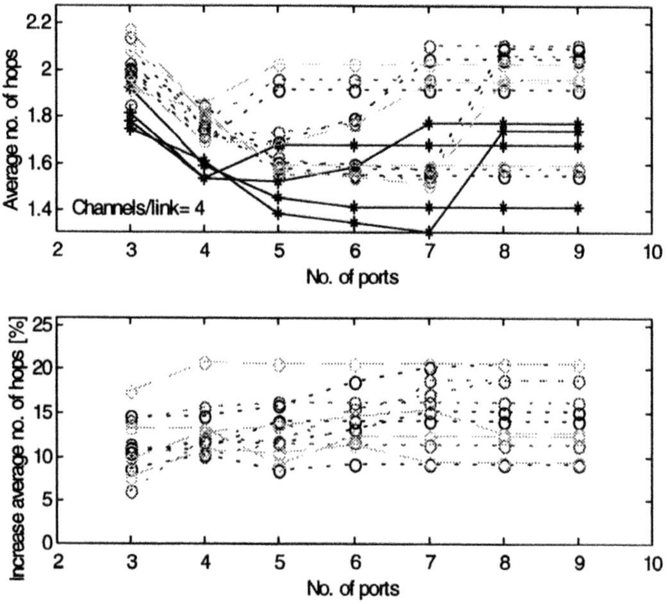

Figure 7 Same as figure 5 but for the 11-node network with 4 WDM-channels per link.

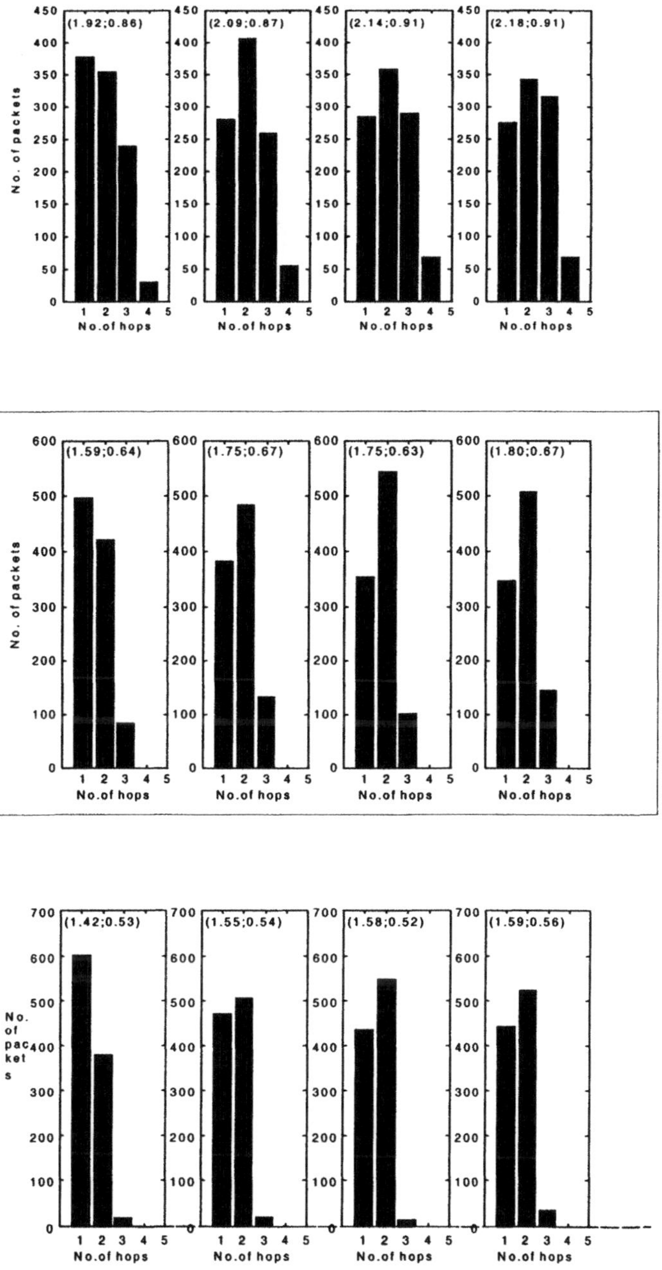

Figure 8 Same as figure 6 but for the 11-node network with 4 WDM-channels per link and 3 ports/node (top), 4 ports/node (middle) and 8 ports/node (bottom).

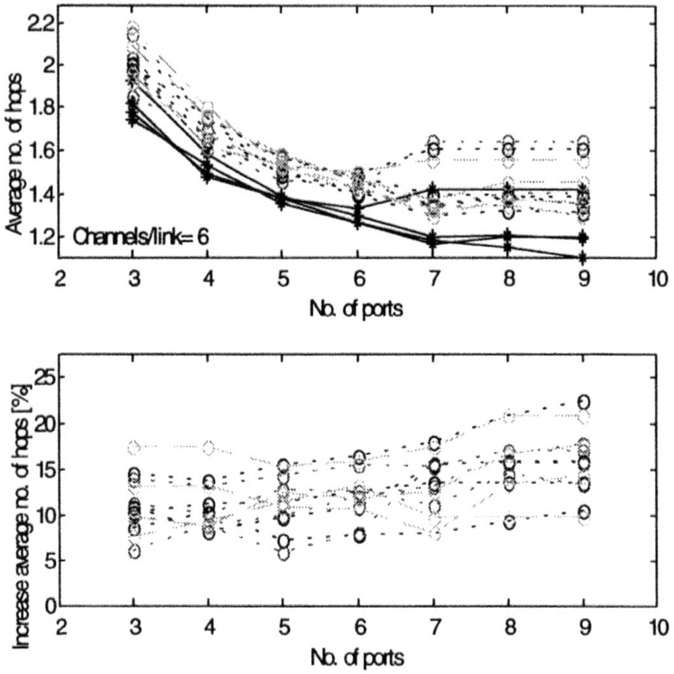

Figure 9 Same as figure 5 but for the 11-node network with 6 WDM-channels per link.

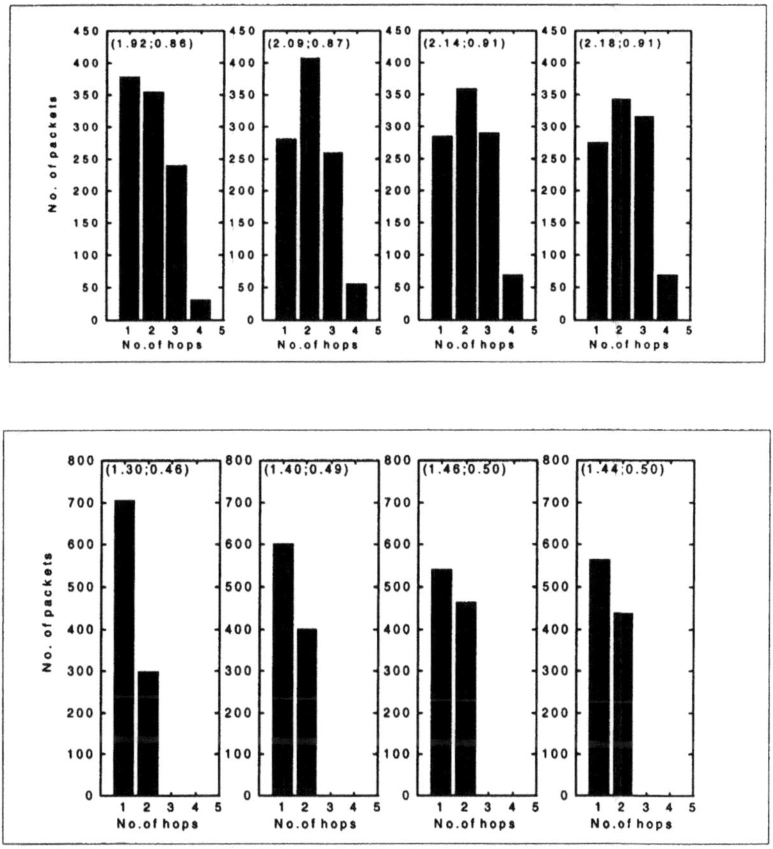

Figure 10 Same as figure 6 but for the 11-node network with 6 WDM-channels per link and 3 ports/node (top) and 6 ports/node (bottom).

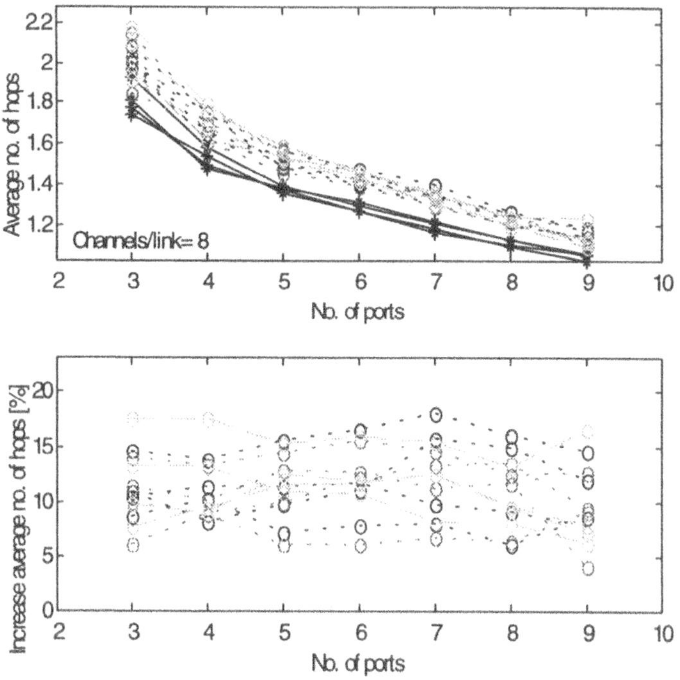

Figure 11 Same as figure 5 but for the 11-node network with 8 WDM-channels per link.

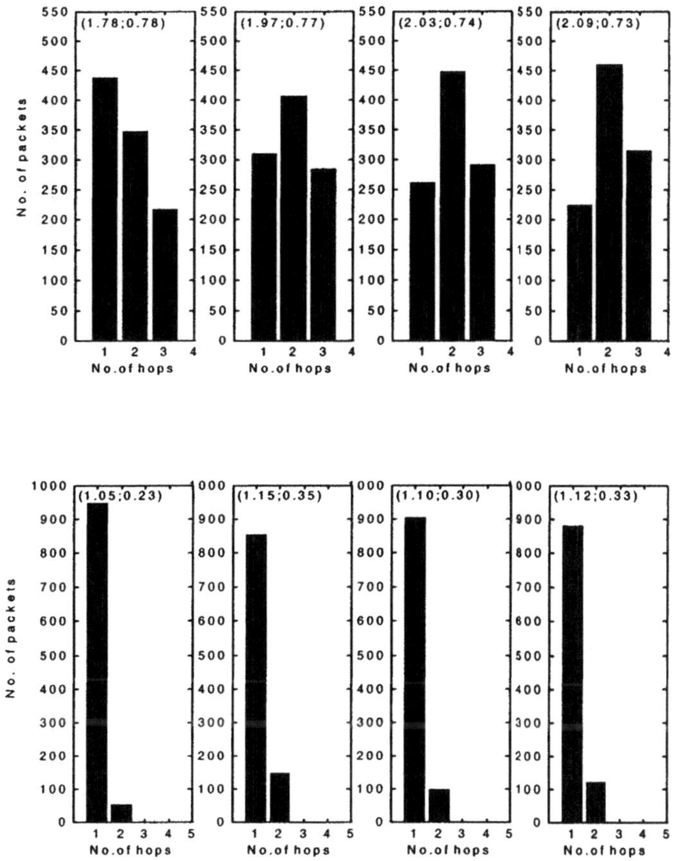

Figure 12 Same as figure 6 but for the 11-node network with 8 WDM-channels per link and 3 ports/node (top) and 9 ports/node (bottom).

As seen in figures 7,9,11, the percentage increase in average number of hops is very dependent on the characteristics of the demand but grows somewhat with the number of ports/node. The average number of hops decrease with the number of ports as expected until the number of ports approaches the number of channels per link, where an increase is evident. The reduced number of hops with more ports is also made clear by figures 8,10 and 12, as well as the benefit of optimising the logical network.

The same calculations are made for the 26-node network, with the results shown in figures 13, 14 for 8 WDM-channels per link, figures 15, 16 for 12 channels per link and figures 17, 18 for 16 channels per link. To assure connected logical networks the minimum number of ports is 4.

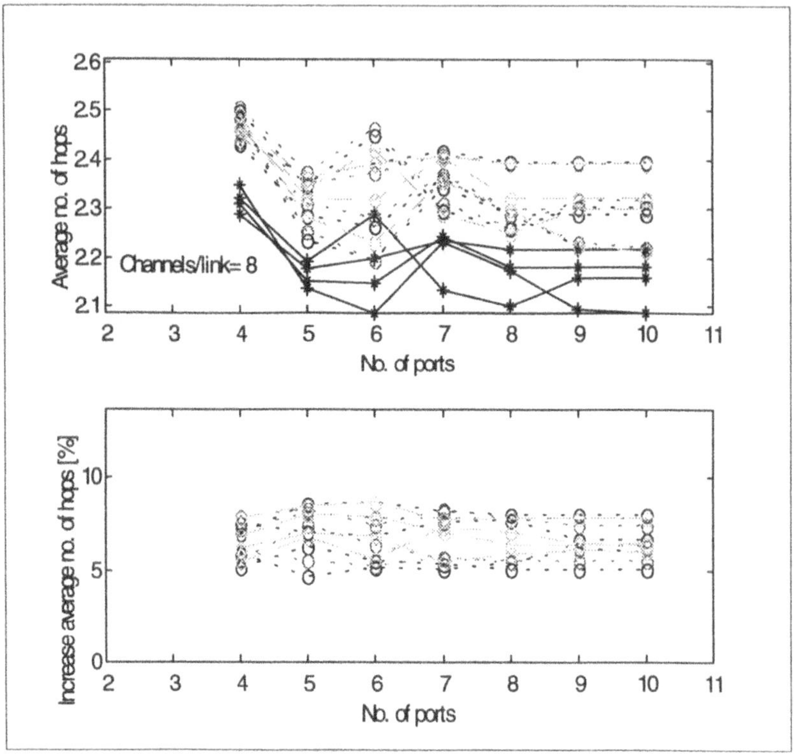

Figure 13 Same as figure 5 but for the 26-node network with 8 WDM-channels per link.

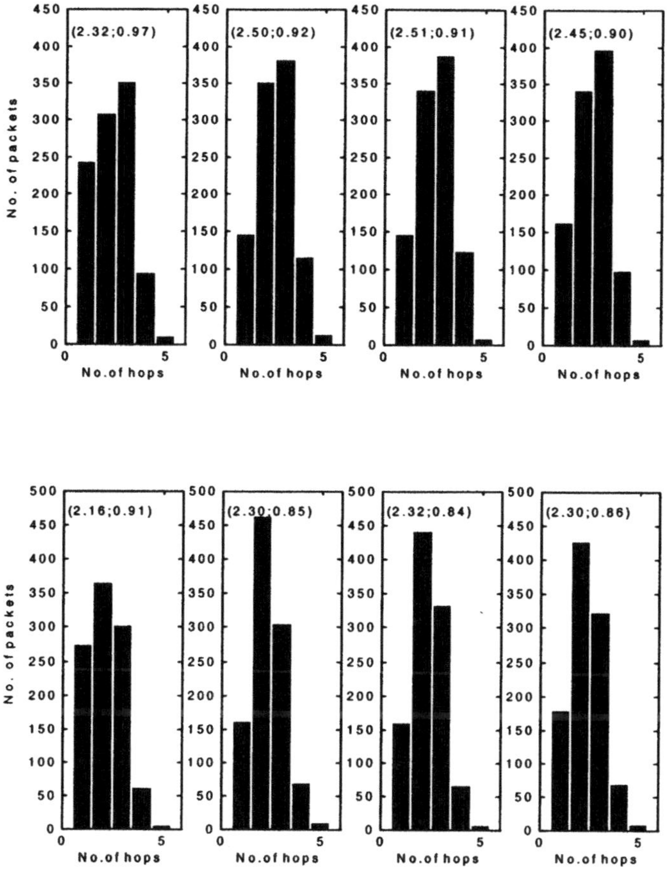

Figure 14 Same as figure 6 but for the 26-node network with 8 WDM-channels per link and 4 ports/node (top) and 10 ports/node (bottom).

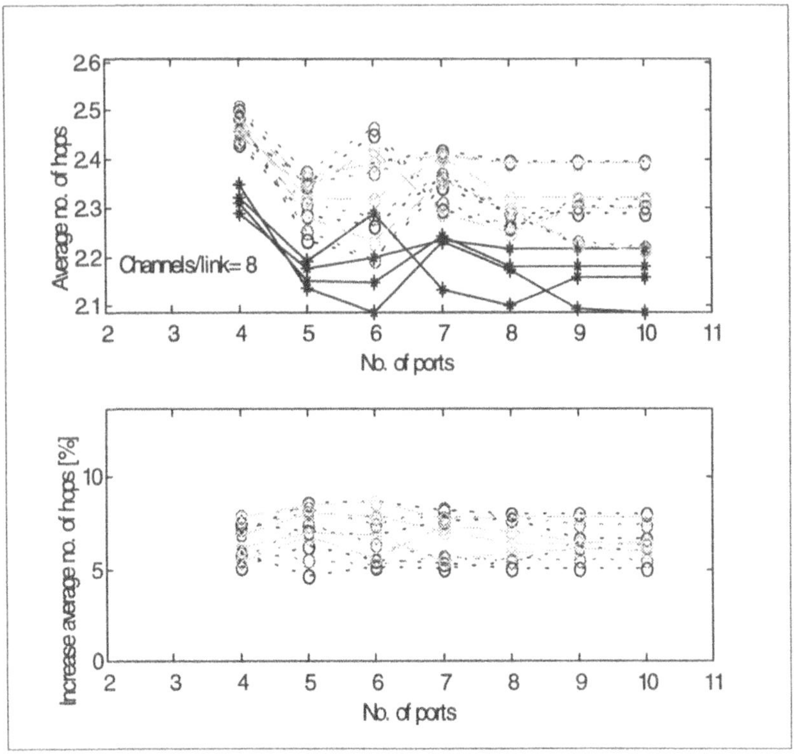

Figure 15 Same as figure 5 but for the 26-node network with 12 WDM-channels per link.

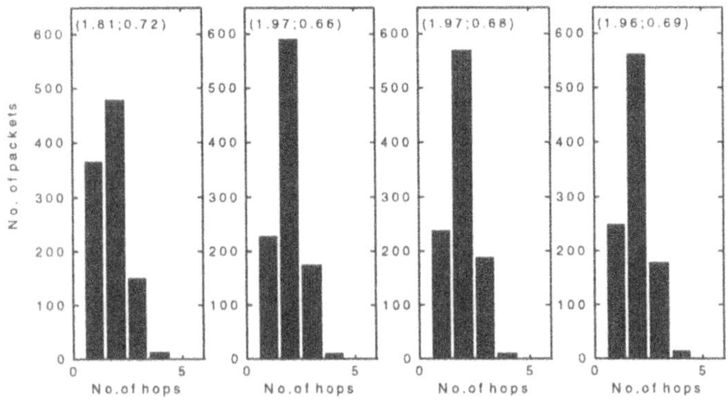

Figure 16 Same as figure 6 but for the 26-node network with 12 WDM-channels per link and 10 ports/node.

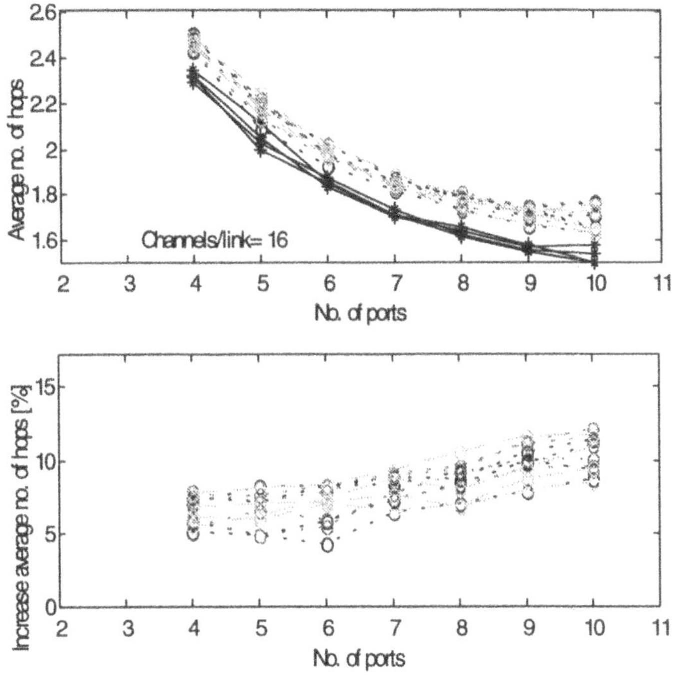

Figure 17 Same as figure 5 but for the 26-node network with 16 WDM-channels per link.

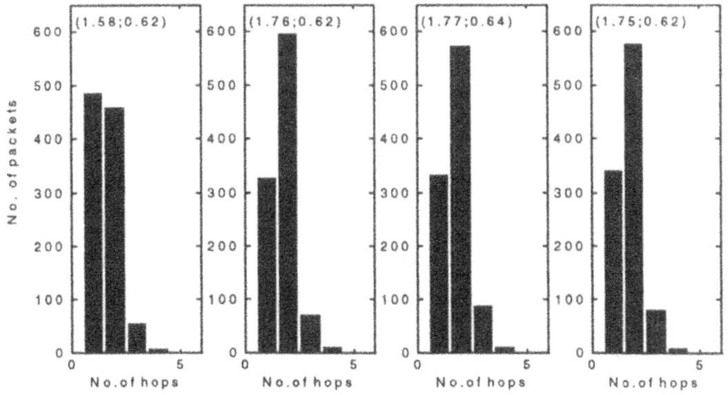

Figure 18 Same as figure 6 but for the 26-node network with 16 WDM-channels per link and 10 ports/node.

As for the 11-node network, the average number of hops decreases with the number of ports but the more complex structure of the 26-node network makes the interaction of ports/node and channels/link less predictable.

However, a low number of channels/link will at some point increase the average number of hops when the number of ports/node increase. The percentage of increase in average number of hops is lower (5-10 %) than for the 11-node network due to longer paths in the 26-node network.

As seen in figure 14, even for the optimised logical networks the average number of hops is larger than 2 due to a low number of ports/node and channels/link. In figures 16 and 18, the average number of hops is lower than 2 due to sufficient number of ports per node and channels per link.

5. DISCUSSION AND CONCLUSION

In this study a number of relevant issues have not been covered and some additional questions remain. A few of them are listed below:

The demands were generated randomly for all node pairs in a uniform way. In real life, however, other types of patterns are common, e.g. server or hub-like patterns where large traffic is carried between a few nodes to/from all other nodes while the traffic between non-hub nodes is negligible. The benefit of wavelength re-configuration in this case is to be further evaluated.

The restriction of a connected network can perhaps be relaxed if re-configuration occurs frequently.

An important factor in determining the gain of having a re-configurable network is the time-scale upon which demands change. (A typical scenario would be the daily change of demand between working areas at daytime and residential areas at evening.) Each network reconfiguration will lead to some loss and misdirection of packets. Loss of packets might occur when cross-connection is done at each OXC while misdirection might occur during connection set-up of a path in the network. The loss of packets in the optical layer therefore depends on the switching time of the OXCs, which is on the order of 10 ns for electrical switch cores, while misdirection of packets depends on the speed of the management system controlling the optical network. Thus, for each network and type of OXC and for the time/volume-characteristics of each traffic demand, an optimal time between network reconfiguration can be expected. Also, by introducing signalling in the network, complexity (=cost) is added and possible extra capacity is needed. Security/reliability in this case is also an interesting issue.

Another issue relating to packet loss and throughput is the need for transfer of information regarding the logical topology state of the optical network to the packet layer /8/.

The reason for re-configuring the whole (sub-)network at once, i.e. having a centralised management, is to avoid possible deadlocks in decentralised connection set-up. For example, if the IP-routers themselves were to control the attached OXC and negotiate with other routers/OXCs, authority over the limited wavelength/port resources is not clear.

This method relies on the possibility of getting statistics on packet numbers and destinations. How this can be realised in practise is not clear.

In conclusion, it has been shown that physical re-configuration can reduce the number of hops in a given physical network. The hop-reduction is in the order of 5-25% depending on the demand variation and increases in general with the number of ports/node. The hop-reduction decreases somewhat with the size of the network.

6. REFERENCES

/1/ R.Ramaswami, K. Sivarajan: "Optical networks", Morgan Kaufmann Publishers, 1998.

/2/ N. Wauters, et al.:"Survivability in new pan-european carriers' network based on WDM and SDH technology: current implementations and future requirements", IEEE Com. Mag., August 1999.

/3/ H. Kobrinski, D. N: Zuckerman:"Applications of automated cross-connection in the optical layer", Mo A2.4, ECOC 1999.

/4/ EURESCOM P918; http://www.eurescom.de/Public/ Projects/p900-series/P918/p918.htm

/5/ E. Goldstein, L.Y.Lin:"Lightwave micromachines for optical crossconnects", Tu A1.1, ECOC1999.

/6/ S.N.Larsson, S. Hubendick, H. Carldén, C. Wickman: "Wavelength label switched IP backbone: architecture and field trial", ThE4 OFC 2000.

/7/ D. Hjelme:"Importance of meshing degree on hardware requirements and capacity utilisation in wavelength routed optical networks", paper 9.2, ONDM'99

/8/ D. Awduche, et al.:"Multi-protocol lambda switching: Combining MPLS traffic engineering control with optical cross-connects", Internet draft, Work in progress, October 1999.

/9/ B. Mukherjee:"Optical communication networks", chapter 9.5, McGraw-Hill 1997.

Resource Allocation for Optical Packet-switching Routers with Self-similar Bursty Traffic

J. J. He and D. Simeonidou
Dept. of Electronic Systems Engineering, University of Essex, Colchester CO4 3SQ, UK

Key words: self-similarity, packet-switching routers, optical buffering, wavelength conversion, multi-path routing

Abstract: This paper addresses the problem of resource allocation in optical packet-switching routers under self-similar bursty Internet traffic conditions. The contention resolution techniques such as optical buffering and wavelength conversion are considered. The router performance is evaluated in terms of packet loss probability and buffer overflow probability via numerical simulations. Results show that wavelength conversion is a promising technique in improving the router performance. Our findings suggest that self-similar traffic significantly increases the resource requirement compared with similar studies performed using traditional bursty traffic models.

1. INTRODUCTION

It is believed that Internet packet traffic will dominate telecommunication network capacity in the near future due to the widely used TCP/IP protocol suite [1]. Consequently packet-switching routers (PSRs) in the network are expected to efficiently handle high throughputs of over 1 Tbit/s. With the development of state-of-the-art optical technology, it is possible to implement the high-throughput PSRs in the optical domain [2].

In designing an optical packet-switching router (OPSR) resource allocation is a critical issue to achieve a good router performance in a cost-effective manner, in terms of packet loss probability (PLP), buffer overflow probability (BOP), packet delay and throughput. Network resources include optical buffers, wavelength converters, tunable lasers, *etc.* Among them the

Optical buffers comprising optical delay lines are crucial and integral parts of the OPSRs. Due to the fact that light travels on-the-fly and can't be removed from the buffer till the end of the buffering period, optical buffers lack flexibility comparing with electronic ones [3]. Moreover the complexity of OPSRs increases with the increase of the number of fiber delay-lines; thus reduction to the number of optical buffers is preferable.

Wavelength converters are essential in improving the efficiency of wavelength-division multiplexed (WDM) networks for both circuit-switching [4] and packet-switching [5] schemes. All-optical wavelength conversion is receiving much attention; the main techniques include four-wave mixing (FWM), difference frequency generation (DFG), cross-gain modulation (XGM) and cross-phase modulation (XPM) in semiconductor optical amplifiers [6]. An ideal wavelength converter should possess such characteristics as transparency to bit rates and signal formats, fast setup time of output wavelength, conversion to both shorter and longer wavelengths, insensitivity to input signal polarization, simple implementation, and so on.

Investigations on resource allocation were reported in the literature [5][7], most of which were based on Markovian assumptions where the packet arrivals were typically modeled by random processes that decay with light-tails, *e.g.* Poisson distribution, leading to short-range dependence (SRD). As a result the router performance can be improved significantly (i.e. exponential decay rate) by slightly increasing the resources. For example the PLP of a router, with an average packet length of 2, a load per channel of 0.4 and 18 buffers, was decreased from 10^{-6} to 10^{-10} with buffers increasing to 32, or to 10^{-11} by employing 2 wavelength conversions [5].

On the contrary, recent studies on traffic traces from packet-based local-area and wide-area networks indicated that Internet traffic is characterized by random process decaying with heavy-tail [8][9]. This results in a long-range dependent (LRD) process that statistically exhibits self-similarity. The self-similar traffic is bursty on a wide range of time scales, a much wider range than that is captured by traditional traffic models.

Since then a lot of work has been carried both on constructing models capable of generating synthetic traffic with matching characteristics with self-similar process [8][9] and on evaluating the effect of self-similar traffic on network performance [10]. Most of the performance evaluations were based on ATM networks, where the cell loss probability and the buffer overflow probability were analytically shown decreasing no more than *hyperbolically* under LRD process but decreasing exponentially under traditional traffic process [10]. These results strongly suggest that there exist necessities to control the self-similarity in the network layer or higher layers and to deal with the self-similar traffic in the physical layer with more careful manner. Recently some of the protocols in the network layer or higher were reported to control the self-similar traffic. In [11] a selective aggressiveness control was adopted for on-line detecting the long-range dependence and consequently improving the throughput. In [12] selected large traffic bursts were shaped at the local access node, which was shown to be capable of decreasing ATM cell loss at a bottleneck queue at the expense of increasing the transmission time of a few large files. However the efficiencies of these methods are to be further refined. On the other hand, there is very little work reported in dealing with self-similar traffic in the lower layers, particularly in the optical packet layer.

In this paper we investigate the resource allocation for optical packet-switching routers conveying self-similar bursty traffic. Optical buffers and wavelength converters are considered as the main network resources. Performance study of OPSRs regarding PLP and BOP is addressed through numerical simulations. In the next section the traffic model and the simulation environment are described. Simulation results and discussions are presented in section III followed by the concluding remarks in section IV.

2. SIMULATION ISSUES

2.1 Traffic model

A process has asymptotic second order self-similarity with parameter $H = (3-\alpha)/2$ if $t^{\alpha}P[\tau > t] \rightarrow z$, as $t \rightarrow \infty$, $1 < \alpha < 2$, (where τ is the active period of traffic and z is a constant) [10]. The larger the value of H, the higher the degree of self-similarity. This characteristic can be well described as a Pareto distribution with cumulative distribution function (CDF) $P[X \le x] = 1-(\theta / x)^{\alpha}$, ($\theta > 0$). When $\alpha \le 1$ it has infinite mean; $\alpha \le 2$, infinite variance; and when $\alpha > 2$, it belongs to non-LRD region [9]. Our investigations will be based on Pareto distributed random process.

2.2 Simulation environment

Traffic process is simulated as a superposition of many sources with Pareto distributed ON and OFF periods [13]. An ON period represents a packet and is always followed by an OFF period for the interval of two consecutive packets. To simplify the simulation, the distribution of ON and OFF periods has the same degree of self-similarity, which equals to having a channel utilisation of 0.5, and are bit padded in integral time-unit lengths suitable for transmission and buffering in optical fibers. The θ in the CDF function signifying the minimum duration generated, is set to 1 in the whole simulation processes. Thus the duration of the ON or OFF period is $\lceil U^{-1/\alpha} \rceil$

in time unit (where U is a random variable uniformly distributed on [0,1] and $\lceil * \rceil$ indicates the ceiling function).

The generic router architecture is shown in Fig.1, which has equal input and output ports with the same wavelength channels. The core of the router is a non-blocking switching network and is controlled by a central control unit not shown in the figure. Each packet is directed to any output port with equal opportunity. For simplicity a FIFO output buffering is considered, which achieves optimal throughput-delay performance compared to other FIFO buffering schemes [7]. Wavelength converters are placed between the output of the router core and the input of the buffers. With output buffering, when two packets compete for the same output wavelength channel, one of them is buffered and will not be transmitted until the channel location is available. If the buffer is full at that particular moment, the packet is lost and the duration of the packet is labeled as buffer overflow moments. With wavelength conversion, this packet is translated to another wavelength channel at the same output port whose buffer or channel is available, thereby resolving the contention and consequently reducing PLP and BOP. All of the simulations were performed over 1×10^6 time units.

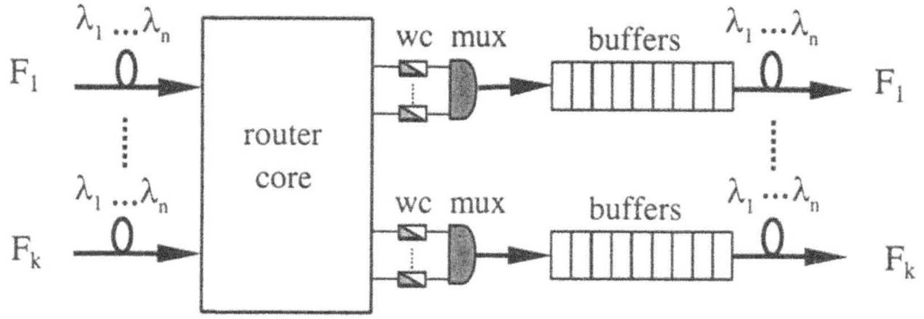

Figure 1. Generic router architecture under simulation

3. RESULTS AND DISCUSSIONS

3.1 Allocation of optical buffers

Firstly resource allocation for OPSRs with optical buffering is studied. In Fig.2 PLP for a 4×4 router (*i.e.* k=4 in Fig. 1) is shown as a function of buffer depth, with α-value being used as a parameter. In non-LRD region (α>2) the decay rate of PLP is exponential, which is in good agreement with the results obtained by using traditional traffic models [5]. While in LRD region (1<α<2) the decreasing rate of the PLP is much less than exponential.

The less the α the heavier the tail of PLP. Thus a large amount of buffers (much more than several tens) will be needed in order to obtain an acceptable PLP of 10^{-10} for self-similar traffic in practical networks.

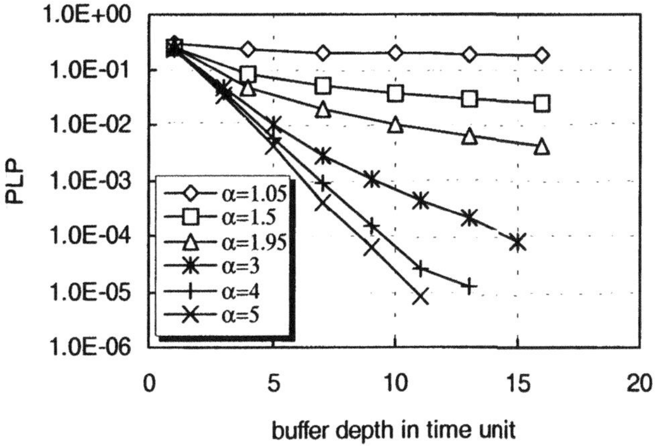

Figure 2. Packet loss probability vs. buffer depth for a 4x4 router with output buffering and different a-values

The same situation exists when investigating the buffer overflow moments in OPSRs. The analytical BOP for the OPSR is developed as follows: [10]

$$P_{overflow} \geq \frac{c_\tau}{\alpha_\tau(\alpha_\tau - 1)(E\tau + E\kappa)^2}\left(\left[\frac{h+b}{a}\right]+2\right)^{-\alpha_\tau+1},$$

where, $c_\tau = \alpha_\tau \theta_\tau^{\alpha_\tau} = \left(\sum_{n=1}^{\infty} n^{-\alpha_\tau - 1}\right)^{-1}$ for ON period distribution $\{\tau_i\}$, and

$E_\tau = c_\tau \sum_{n=1}^{\infty} n^{-\alpha_\tau}$; $c_\kappa = \alpha_\kappa \theta_\kappa^{\alpha_\kappa} = \left(\sum_{n=1}^{\infty} n^{-\alpha_\kappa - 1}\right)^{-1}$ for OFF period distribution $\{\kappa_i\}$,

and $E_\kappa = c_\kappa \sum_{n=1}^{\infty} n^{-\alpha_\kappa}$; $a = \frac{1}{E_\tau + E_\kappa}$, $b = a+1$, and h is the buffer depth for a wavelength channel.

In Fig.3 the analytical lower bound and the simulation results of BOP are plotted as a function of buffer depth with the number of ports as a parameter for the middle degree of self-similarity (α=1.5). As can be seen that both of the analytical and simulated results of BOP do not decrease faster than hyperbolically with the increase of buffer depth for different routers. In addition, BOP increases but the increase rate slows down as the number of port increases. Analytical results show that 10^9 buffers will be needed to obtain a BOP lower than 10^{-5} for $\alpha=1.5$.

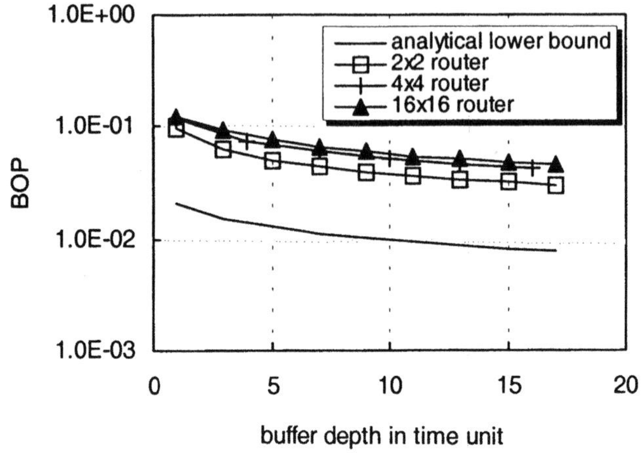

Figure 3. Comparison between simulated and analytical overflow probability for routers with output buffering and $\alpha=1.5$

3.2 Allocation of wavelength conversions

Next the resource allocation for routers with wavelength conversion (WC) and optical buffering is investigated. Fig.4 shows PLP as a function of buffer depth for 4×4 router with self-similar traffic of $\alpha=1.5$. It can be observed that PLP decreases with the deployment of WC. Moreover the use of WC changes the heavy tail distribution of the PLP, which shows itself a promising resource for the future OPSRs.

To further verify this point the PLP versus the number of WCs is depicted in Fig. 5 for a 4×4 router with 16, 50, 100, 300 buffers respectively. The super-linear projection lines corresponding to different amount of buffer depth are plotted in Fig. 5 as well. As can be seen that PLP decreases two orders in magnitude for the case of 6 WCs and 50 buffers comparing with the case of 300 buffers only, and decreases one order in magnitude comparing with that of 3 WCs and 100 buffers. It can be expected that as the number of WCs increases the PLP decreases. The rate of decrease strongly depends on the buffer depth. As evidence, with 300 buffers the PLP decreases much faster than that with 16 buffers. The projection line for the case of 300 buffers indicates that to obtain the PLP of 10^{-10} there will need 12 wavelength conversions. Compared with traditional traffic models, the resources required under LRD traffic model are much greater than the case of traditional SRD traffic.

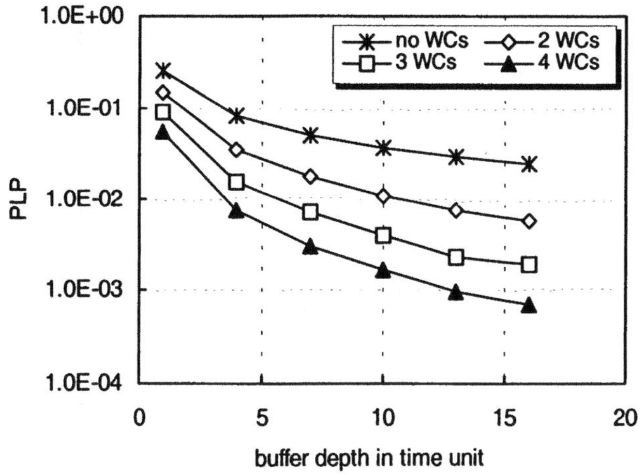

Figure 4. Packet loss probability vs. buffer depth for a 4×4 router with wavelength conversion and α=1.5

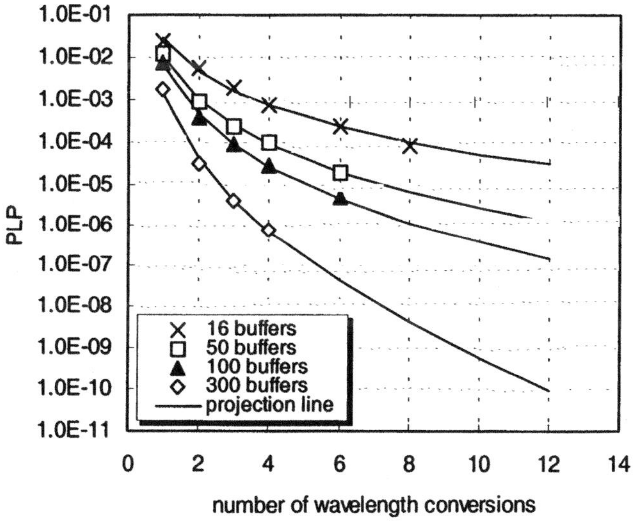

Figure 5. Packet loss probability vs. number of wavelength conversions for a 4×4 router with different buffer depth and α=1.5

4. CONCLUSIONS

A model describing the self-similarity of Internet traffic has been used in investigating the resource allocation for optical packet-switching routers. The router performance has been evaluated through simulations. It has been shown that optical buffering on its own could not offer adequate router performance. The addition of wavelength conversions resulted in a decrease of the heavy tail distribution of packet loss probability and thus a better router performance. Our work strongly suggests that in comparison with traditional traffic models the self-similar traffic does heavily influence the resource requirement in the router.

Acknowledgements: The authors wish to thank Corning Optical Fibres for funding this research project.

References

[1] S. A. Thomas, Ipng and the TCP/IP protocols implementing the next generation Internet, John Wiley & Sons, Inc., 1996

[2] M. Renaud, F. Masetti, C. Guillemot and B. Bostica, "Network and system concepts for optical packet switching", *IEEE Communications Magazine*, April 1997, pp. 96-102

[3] D.K. Hunter, M.C. Chia and I. Andonovic, "Buffering in optical packet switches", *Journal of Lightwave Technology*, 1998, vol. 16, pp. 2081-2094

[4] R.A. Barry and P.A. Humblet, "Models of blocking probability in all-optical networks with and without wavelength changers", *IEEE Journal on Selected Areas in Communications,* 1996, Vol. 14, pp. 858-867

[5] S. L. Danielsen, C. Joergensen, B. Mikkelsen, and K. E. Stubkjaer, "Analysis of a WDM packet switch with improved performance under bursty traffic conditions due to tuneable wavelength converters", *Journal of Lightwave Technology*, 1998, vol. 16, pp. 729-735

[6] B. Ranamurthy and B. Mukherjee, "Wavelength conversion in WDM networking", *IEEE Journal on Selected Areas in Communications*, 1998, vol. 16, pp. 1061-1073

[7] M. G. Hluchyj and M. J. Karol, "Queueing in high-performance packet switching", *IEEE Journal on Selected Areas in Communications*, 1988, vol. 6, pp. 1587-1597

[8] W. E. Leland, M. S. Taqqu, W. Willinger and D. V. Wilson, "On the self-similar nature of Ethernet traffic (extended version)", *IEEE/ACM Transactions on Networking*, 1994, vol. 2, pp. 1-15

[9] V. Paxson and S. Floyd, "Wide area traffic: the failure of Poisson modeling", *IEEE/ACM Transactions on Networking*, 1995, vol. 3, pp. 226-244

[10] B. Tsybakov and N. D. Georganas, "On self-similar traffic in ATM queues: definition, overflow probability bound and cell delay distribution", *IEEE/ACM Transactions on Networking*, 1997, vol. 5, pp. 397-408

[11] T. Tuan and K. Park, "Multiple time scale congestion control for self-similar network Traffic", *Performance Evaluation*, 1999, vol. 36~7, pp. 359-386

[12] K.J. Christensen and V. Ballingam, "Reduction of self-similarity by application-level traffic shaping", 22nd *IEEE Conference on Local Computer Networks*, pp. 511-518, Nov. 1997

[13] L. S. Tamil, F. Masetti, T. McDermott, G. Castanon, An Ge and L. Tančevski, "Analysis of all-optical IP routers", *SPIE Conference on All-optical Networking*, pp. 456-464, Boston, Nov. 1998

[14] G. A. Castañón, O. K. Tonguz and A. Bononi, "Benefits of wavelength translation in datagram all optical networks", *Electronics Letters*, 1997, vol. 33, pp. 1567-1568

A Flexible WDM Ring Network Design and Dimensioning Methodology

A. Lardies[1], R. Jagannathan[1], A. Fumagalli[2], I. Cerutti[2], M. Tacca[2]
[1] *Alcatel Corporate Research Center*
1201, E.Campbell Road,
Richardson, TX – 75081, USA
ana.lardies@aud.alcatel.com

[2] *University of Texas at Dallas*
CATSS: Center for Advanced Telecommunications Systems and Services
P.O. Box 830688 – Richardson, TX 75083
e-mail:{andreaf, isabella, mtacca}@utdallas.edu

Abstract: Designing arbitrary topologies based on Wavelength Division Multiplexing (WDM) rings yields a number of advantages, including fast restoration mechanisms, simple network management and control. The optimal design of such networks is however a rather complex task, with multiple objectives being optimized that can often be in conflict one with another. The paper presents a modular methodology that encompasses three fundamental steps in designing an arbitrary topology organized as a set of interconnected WDM rings. These are the selection of the Ring Cover, the Routing of the inter-ring traffic and the Traffic Balancing with Wavelength Assignment (TBWA) of the traffic locally on individual rings. After a description of the overall methodology, a detailed formulation of the TBWA problem in presence of both symmetric and asymmetric traffic demands is given. An efficient algorithm that yields sub-optimal solution of the TBWA problem is then proposed. Numerical results show that the algorithm finds an optimum solution in a number of experiments.

1. INTRODUCTION

Wavelength Division Multiplexing (WDM) technology is fast becoming the de facto standard solution to the user's continuously growing demand for high bandwidth connections. At the same time, various customers nowadays demand telecommunication networks that are capable of providing highly reliable and uninterrupted service. Combined together, these two facts are creating an increasing interest in the design of survivable WDM backbones with built-in protection mechanisms available in the optical layer. In this work we focus our attention on ring-based protection mechanisms as they are often used in voice and data networks [10].

A ring based network design yields a number of advantages, including simplified management and control of the network, and fast restoration mechanisms upon link or node failure occurrence. The design of a given arbitrary topology in the form of a set of interconnected rings reduces the overall network connectivity and hence may limit the number of distinct cables that carry working fibers in the final network layout. With a reduced number of cables in the system the overall network cost may also be reduced. Due to these and other advantages, it is not surprising that ring-based network designs are often preferred to other network topologies, e.g., meshed topologies, even when these alternative solutions can statistically save up to 30-60% of the spare resources.

It is envisioned that in WDM networks too, a ring-based design shall offer advantages similar to those that are found in conventional networks. Considering that in a WDM network, the abundant bandwidth is not an expensive, nor a scarce resource, the 30-60% spare resource savings of mesh layouts may become even less critical from an overall network design viewpoint.

Multiple criteria can be defined to optimally design and dimension WDM ring-based networks. In a previous work, the authors considered the problem of minimizing the total wavelength mileage [1], defined as the sum of the working and protection wavelength miles required in the network to support a given set of traffic demands. In [1], an exact Integer Linear Programming (ILP) formulation of the problem and a simplified ILP formulation that can handle the design of large networks were proposed. In a more recent work [2] the problem of minimizing the total number of optical cross-connects ports in the network was addressed and a sub-optimal solution based on a greedy algorithm was proposed.

This paper addresses the problem of minimizing the total fiber mileage required in a given network with arbitrary topology to support a given set of traffic demands. The problem is solved assuming that wavelength converters are not available for intra-ring traffic and traffic demands are either symmetric (e.g., voice traffic) or asymmetric (e.g., client-server data traffic). The problem is defined assuming that SONET (2-fiber Bidirectional Line Switched Rings) 2f-BLSR like optical rings [10] are used to design the network. In order to design a network that is reliable in presence of either a link or a node fault, inter-ring traffic is supported through dual homing, i.e., at least two cross-connects exist that can interconnect the traffic demands between two adjacent rings.

The total fiber mileage in the network is reduced using a modular approach that consists of three fundamental steps. These are the selection of the ring cover, the routing of the inter-ring traffic and the Traffic Balancing with Wavelength Assignment (TBWA) of the traffic on individual rings. The paper briefly describes the first two steps, it then gives a detailed formulation of the TBWA problem in presence of both symmetric and asymmetric traffic demands. An efficient algorithm that yields sub-optimal solution of the TBWA problem is then presented and numerical results for a particular network configuration are shown, demonstrating that, in spite of its limited complexity ($O(F \cdot N_r^2)$ where F is the maximum number of fibers per link, and N_r is the number of nodes in the ring), the proposed algorithm reaches an optimum solution in a number of experiments.

Although only the solution of the TBWA problem is addressed in detail in this work, it must be noticed that due to its modularity, the proposed design and dimensioning methodology can be easily adjusted to optimize various objective functions. It can thus provides the designer with a fairly efficient and versatile optimization technique to design optical networks with ring-based protection schemes.

2. PLANNING METHODOLOGY

The presented approach consists of splitting the overall optimization problem into a number of sub-problems that can be handled separately and solved sequentially. On one hand, it is well known that dividing the optimization problem into sequential sub-problems may not result in an optimum solution when dealing with a specific cost function, e.g., total wavelength mileage [1]. On the other hand it offers a significant simplification of the optimization process. Consequently, more system parameters can be taken into account during the design optimization, or multiple solutions to the problem can be found in a short time interval. In addition, it yields a modular solution that enhance the versatility of the designed optimization algorithms that can thus accommodate different planning optimization criteria, and satisfy the various, possibly changing over time, designer's needs.

The planning methodology proposed in the paper consists of three steps: the ring cover of the arbitrary topology, the routing of the traffic demands between rings minimizing the number of crossconnections, and the TBWA in each ring. The variables under optimization can be modified in any of these steps to improve the final outcome of the overall design optimization.

A ring cover is defined as a set of rings such that the union of all their spans constitutes the network topology itself. Thus, a ring cover overlaid on a network topology ensures full connectivity of the network nodes. The first step is to identify the ring cover from a set of candidate rings $\boldsymbol{R}$. Having an independent set of candidate rings gives flexibility to the designer to include or exclude a certain set of rings in the design optimization. If the designer has already selected a cover for the network, that cover is then entered as $\boldsymbol{R}$ and the first optimization step is skipped altogether. In the second step every traffic demand is routed over the ring cover from the source node to the destination node. This step determines which rings a demand passes through. This step allows the designer to choose the routes (in terms of sequence of rings) either by means of some standard routing algorithm or manually. In the third step the traffic demands assigned to each ring are optimally routed (clockwise or counter clockwise) and assigned a wavelength within the ring. In this step the λ granularity of each fiber, i.e., the number of wavelengths per fiber, is taken into account as an input parameter.

Each step of the proposed planning methodology, with particular attention to the TBWA problem, is described next.

2.1 Ring Cover

First, an initial (possibly large) set of candidate rings $\{\boldsymbol{R}_1, \boldsymbol{R}_2, \ldots\}$ is generated according to the minimum and maximum number of nodes allowed in each ring. A ring cover is formally defined as a selection of rings $C = \{R_{n_1}, \ldots, R_{n_k}\}$, subset of the candidate rings, such that all nodes generating or terminating traffic belongs to at least one ring in $\boldsymbol{C}$. In addition, any ring $\boldsymbol{R}_x \in \boldsymbol{C}$ must have at least two nodes $\boldsymbol{i}, \boldsymbol{j}$ such that there is at least one other ring $\boldsymbol{R}_y \in \boldsymbol{C}$ and $\boldsymbol{i}, \boldsymbol{j} \in \boldsymbol{R}_y$. This second condition ensures that no ring in the cover is isolated, and any ring is connected to at least another ring in the cover via at least two nodes. With the above cover, traffic can be routed in such a way that any single node or link fault will not disrupt any connection. The ring interconnection can be protected with a *drop-and-continue* mechanism, or by duplication of the Network Element (NE) components. Considering the high node availability when compared to the link availability, and the fact that the *drop-and-continue* mechanism requires additional bandwidth in the rings, the solution proposed here is based on duplication of components in the ring interconnection nodes. Consequently, the scope of the paper is limited to the case of (single) link failure. Dual homing between adjacent rings may be still required to allow traffic balancing of the inter-ring traffic between the two or more cross-connects that interconnect the rings.

Many objectives can be identified while searching for an optimal ring cover, including the minimization of:

- Total fiber mileage
- Total cable mileage
- Total wavelength mileage
- Number of active spans
- Number of optical cross connects (OXC)
- Number of OXC ports
- Size of biggest OXC etc.

In this work the ring cover is selected with the objective to minimize the sum of the perimeters of the selected rings.

2.2 Inter-Ring Routing

Once the ring cover is determined the next step is to route each traffic demands across the rings, e.g., see [9]. This step confines itself to identifying the sequence of rings that a traffic demand is routed through, but not the exact physical path for the working connection. Using the shortest path algorithm it is possible to route the demands with the objective of obtaining the shortest number of rings crossed, or the smallest sum of ring perimeters, or the shortest physical path etc. In the paper the routing of the traffic demands is obtained with the objective to minimize the number of rings crossed by every demand. The routing of the traffic demands may also be done according to some additional engineering rules that are provided by the designer and that are external to this design step, e.g. availability of wavelength converters. At the end of this step every demand must be mapped into a sequence of rings, and an initial selection of the nodes cross-connecting each demand between adjacent rings has been made. Once the inter-ring routing has been determined, the traffic demands within each individual ring can be created by determining the ingress and egress points of each demand in the ring.

2.3 Routing and Wavelength Assignment in the WDM Ring

The third and last step in the design process is to find, within each WDM ring, an optimal routing and an optimal wavelength assignment. As already mentioned, among the several SONET-like optical ring architectures, the 2f-BLSR architecture is considered in the paper. In a bidirectional ring the optical channels are transmitted in both directions of the ring (clockwise and counter-clockwise), and the corresponding protection connections are routed with opposite direction. Line protection mechanism is considered. The spare bandwidth can be shared to protect several connections. Due to the fact that all the links on the ring have the same capacity, the actual capacity of the ring is determined by the capacity of the link with highest traffic load. If the optical channels are symmetric, then for every node pair *(i, j)* in the ring, the channels from *i* to *j* and the channels from *j* to *i* must use the same route. It is assumed that the number of wavelengths in the fiber is fixed and half of these wavelengths are dedicated to working connections, while the others are dedicated to protection. Wavelength conversion is allowed only at the nodes that interconnect traffic demands between rings, and only for the inter-rings traffic demands.

The objective of this step is to determine the route for the traffic demands in the ring and the working and protection wavelengths that reduce the number of fibers in the ring.

This problem can be demonstrated to be NP-hard, based on the demonstration given in [4] for a similar problem. In this section we will define the mathematical formulation of this problem. We also present a heuristic algorithm which solves this problem in $O(F \cdot N_r^2)$ time, where F is the number of fibers per link, and N_r is the number of nodes in the ring.

2.3.1 Problem Formulation for Symmetric Connections

Given a graph $G(N,E)$ representing the arbitrary topology under design optimization, we consider each ring separately. We define a graph representing each ring R as $R(N_r, E_r)$ where N_r is the set of nodes in ring R, and E_r is the set of spans in ring R.

We assume that the channel protecting a working channel must use the same wavelength assigned to the working channel. As a result, wavelength conversion between the working and the protection channels is not required when a fault occurs. Furthermore we assume that bidirectional line shared switched ring architecture is used, therefore, wavelength channels dedicated to protection can be shared among a number of traffic demands. Based on these assumptions, we formulate the following a Integer Linear Problem (ILP) formulation that minimizes the number of fibers necessary in the ring to fulfill the traffic requests and ensure 100% protection against any single link fault.

Given:

W : Maximum number of wavelength per fiber

d_{ij} : Number of bidirectional optical channels in ring R between source destination pair i,j, (sum of both intra-ring traffic and inter-ring traffic)

$P_d(i,j)$: Available path in direction d of the ring for the connection from node i to node j, where d is 1 for the clockwise direction, and 2 for the counter clockwise direction.

L_d : Set of links in the ring oriented along direction d, obviously $L_1 \bigcup L_2 = E_r$

Decision variables:

E_{ij}^{wfd} : 1 if source destination pair i,j is assigned wavelength w in fiber f in direction d of the ring

0 otherwise

L^{wfd} : 1 if wavelength w in fiber f, is assigned to some connection routed in

direction d of the ring

0 otherwise

P^{wfd} : 1 if wavelength w in fiber f in direction d of the ring is necessary

for protection

0 otherwise

F^{fd} : 1 if fiber f is in direction d of the ring is used, either for working connections

or for protection channels

0 otherwise

The problem can be formulated as follows:

$$\textit{Minimize} \quad \sum_f \sum_d F^{fd}$$

Subject to:

$$W \cdot F^{fd} \geq \sum_w (P^{wfd} + L^{wfd}), \quad d = \{1,2\} \;\; \forall f \qquad (1)$$

$$\sum_w \sum_f (E_{ij}^{wf1} + E_{ij}^{wf2}) = d_{ij} \qquad \forall i,j \in N_r \qquad (2)$$

$$\sum_{\{i,j \in N_r\} | l \in P_d(i,j)} E_{ij}^{wfd} \leq 1 \quad \forall l \in L_d, \forall f, \forall w, d = \{1,2\} \qquad (3)$$

$$\sum_{i,j\in N_r} E_{ij}^{wfd} \leq K \cdot L^{wfd} \quad d=\{1,2\}\ \forall f \ \ \forall w \tag{4}$$

$$P^{wf2} = L^{wf1},\ P^{wf1} = L^{wf2} \quad \forall w,\ \ \forall f \tag{5}$$

$$L^{wfd} + P^{wfd} \leq 1 \quad \forall f \ \ \forall w,\ d=\{1,2\} \tag{6}$$

$$\sum_{w}\sum_{f} E_{ij}^{wf1} = \sum_{w}\sum_{f} E_{ji}^{wf2}, \qquad \forall i,j \in N \tag{7}$$

$$E_{ij}^{wf1},\ L^{wfd},\ P^{wfd},\ F^{fd} \in \{0,1\}$$

with K >0, big enough.

The objective is to minimize the number of fibers in the ring. Restriction **(1)** permits to use at most W wavelengths for working or protection purposes in fiber f and direction d. Restriction **(2)** guarantees that the offered traffic demands are fulfilled. Restriction **(3)** ensures that on each link, in each direction, on each fiber, a wavelength can be assigned to at most 1 connection. Restriction **(4)** guarantees that if any working connection is using wavelength w in fiber f and in direction d, then this wavelength has to be assigned for working traffic and cannot be used for protection. Restriction **(5)** guarantees for each assigned working wavelength, the existence of a protection wavelength on the counter rotating fiber. Restriction **(6)** guarantees that every wavelength in each fiber is assigned to protection or working connections but not to both of them. Restriction **(7)** guarantees the use of the same route for bidirectional connection demands.

2.3.2 Heuristic Algorithm for Symmetric Connections

The algorithm comprises five steps. The routing of the demands in the ring is solved in the first four steps, the wavelengths are assigned in the last one.

Step 1. Initial Routing

Each demand is routed through the ring minimizing the number of hops.

NC_{ij}^{l} : Number of optical channels between source destination pair i-j that are crossing link l, $i,j\in N_r$, $l\in E_r$.

Step 2. Determine the Link Load

For each link, calculate the number of channels per link:

$$CL^{l} = \sum_{i,j\in N_r} NC_{ij}^{l}$$

The link l^* with highest traffic load is calculated as follows:

$$CL^{l^*} = \max_{l \in E_r}\{CL^l\} = \text{CL}$$

Step 3. Optimize the Routing

The objective of the algorithm is to reduce CL. This maximum is reached on link l^*, $l^* \in E_r$. Then, some demands routed through l^* should be re-routed in the opposite direction of the ring.

$S = \{(i,j) / i, j \in N_r, NC_{ij}^{l^*} > 0\}$ set of demands that could be re-routed.

Repeat for each source destination pair $i^*\text{-}j^* \in S$

L'={l∈L | link l will be crossed if the connection (i*,j*) is re-routed}

$$Max2 = \max_{l \in L'}\left\{\sum_{i,j \in N_r} NC_{ij}^{l}\right\},$$

Number of channels i*-j* to be re-routed:

$$\Delta(i^*, j^*) = \min\left\{\tfrac{1}{2}(CL - Max2), NC_{i^* j^*}^{l^*}\right\}$$

It is easy to show that this algorithm solves the routing problem in $O(N_r^2)$ time, where N_r is the number of nodes in the ring.

Step 4. Selection of the Node Interconnection

For each external connection to the ring it is necessary to decide the interconnection node. An initial assignment was done in a previous module (Sec. 2.2), while mapping the traffic demands over the rings selected for the cover. After step 3 some working connections may have been re-routed along the ring, this has a direct impact on the OXC node selection. The OXC node selection, i.e. the ring interconnection nodes, for each demand may have to be reselected in order to guarantee that after balancing the total demand length, in terms of number of hops, is still minimal, eliminating the loops between OXC that could be produced by the balancing algorithm in each ring.

Step 5. Wavelength assignment

When the optimal routing has been obtained, these demands have to be assigned to one wavelength minimizing the number of fibers in the ring.
It is not possible to evaluate the exact number of wavelengths necessary in each ring, because wavelength conversion is not permitted, however it is possible to find a lower bound and an upper bound. The maximum number of working wavelength necessary in each direction is:

$$nww_{\max} = 2 \cdot CL - 1$$

and the minimum number is:

$$nww_{\min} = CL$$

Then ws assign each demand to a wavelength along the ring, we expect a number of working wavelengths, in each direction of propagation, to be:

$$nww_{\min} \leq nww \leq nww_{\max}$$

Then we calculate the number of fiber pairs necessaries in each link of the ring as:

$m = \left\lceil \frac{2 \cdot nww}{W} \right\rceil$, where W is the number of wavelength per fiber.

Demands are uniformly distributed among these fibers. In each fiber half of the wavelength are assigned to working connections and the other half to protection channels. This TBWA problem is solved in $O(F \cdot N_r^2)$ time, where F is the number of fibers per link

2.3.3 Problem Formulation for asymmetric connections

The formulation is the same one used for the symmetric connections, except that restriction (7) is not necessary due to the connection asymmetry.

2.3.4 Heuristic Algorithm for asymmetric connections

Step 1. Initial Routing

The same as the ones defined for symmetric traffic

Step 2. Determine the Link Loading

Same as that defined for symmetric traffic, but calculated for each direction of the ring:

$$CL^{l_1} = \sum_{i,j \in N_r | l \in P_1(i,j)} NC_{ij}^{l} \qquad \forall\, l \in L_1$$

$$CL^{l_2} = \sum_{i,j \in N_r | l \in P_2(i,j)} NC_{ij}^{l} \qquad \forall\, l \in L_2$$

The links l_1^*,and l_2^* with highest traffic load will be calculated as follows:

$$CL^{l_1^*} = \max_{l \in E_r}\left\{CL^{l_1}\right\} \quad CL^{l_2^*} = \max_{l \in E_r}\left\{CL^{l_2}\right\}$$

Step 3. Optimize the Routing

The objective of the algorithm is to minimize $CL = \max\left\{CL^{l^*1}, CL^{l^*2}\right\}$.

This maximum is reached in the link **l***, $l^* \in E_r$. Then, some demands routed through l_1^*, l_2^* should be re-routed in the other direction of the ring. The following steps are repeated until no more connection can be re-routed, and then go to step 4.

- If $CL = CL^{l^*1}$, then $S^{l_1^*} = \left\{(i,j) / i, j \in N_r, NC_{ij}^{l_1^*} > 0\right\}$ set of demands that could be re-routed. Sort this set in decreasing order of the number of links crossed for each demand. Repeat for each pair $(i^*,j^*) \in S^{l_1^*}$
  L'={l∈L / link l will be crossed if the connection (i*,j*) is re-routed}
  $$Max2 = \max_{l \in L'}\left\{\sum_{i,j \in N_r} NC_{ij}^{l}\right\},$$
  Number of channels i*-j* to be re-routed:
  $$\Delta(i^*, j^*) = \min\left\{\left\lfloor \frac{CL - Max2}{2} \right\rfloor, NC_{i^*j^*}^{l^*}\right\}$$
- If $CL = CL^{l^*2}$, the process is similar to the previous one when $CL = CL^{l^*1}$, but with the other direction of the ring

Step 4. Selection of the Node Interconnection
The same as the one defined for symmetric traffic

Step 5. Wavelength assignment
The same as the one defined for symmetric traffic

3. NUMERICAL RESULTS

A long distance network has been selected in order to apply the above methodology.

Thc initial sct of rings comprises 170 rings and the result obtained by ring cover optimization, as depicted in the figure 1, is formed by only 10 rings.

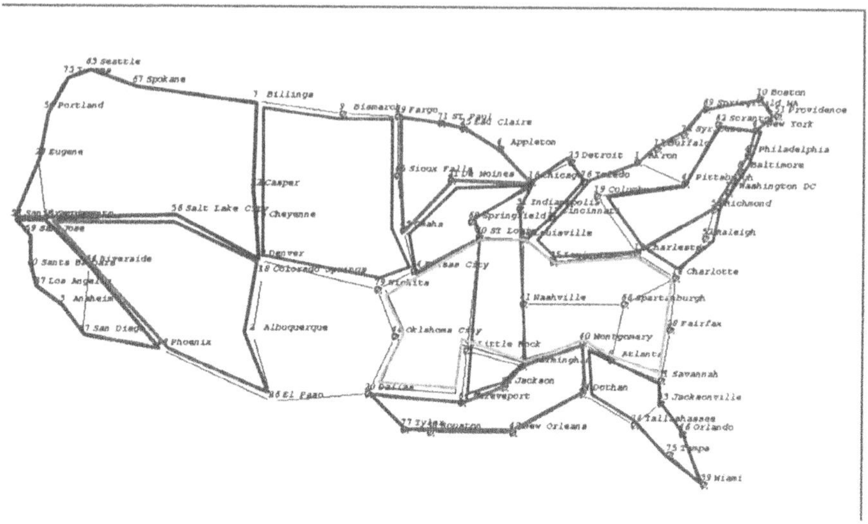

Figure 1. Ring cover result

We have applied the presented design methodology to the network in Figure 1, assuming asymmetric traffic demands. In order to illustrate the behavior of the TBA, the network has been also dimensioned without balancing the traffic in the rings. The main results are represented in the Table 1. As we can see after the Traffic Balancing the fiber length has been reduced around 20%, and around 7% the total wavelength mileage.

In Figure 2, for each link in the network the necessary number of fibers calculated by the proposed heuristic is reported, with and without traffic balancing. After the traffic balancing more than 15% of the links need 6 fibers. Less than 5% of the links need 4 fibers. As indicated in the figure, around 6% of the links do not need any fiber, i.e., the links are not covered. This is an acceptable result since the objective of the design is to protect all the generated demands, and not necessarily all the links of the arbitrary topology.

	No Traffic Balancing Algorithm	**After Traffic Balancing Algorithm**
Fiber Length	200,276	247,613
Working Wavelength miles	616,487	535,963
Protection Wavelength miles	1,310,450	1,541,330
Total Wavelength miles	1,926,937	2,077,293

Table 1. General Ring Dimensioning Results

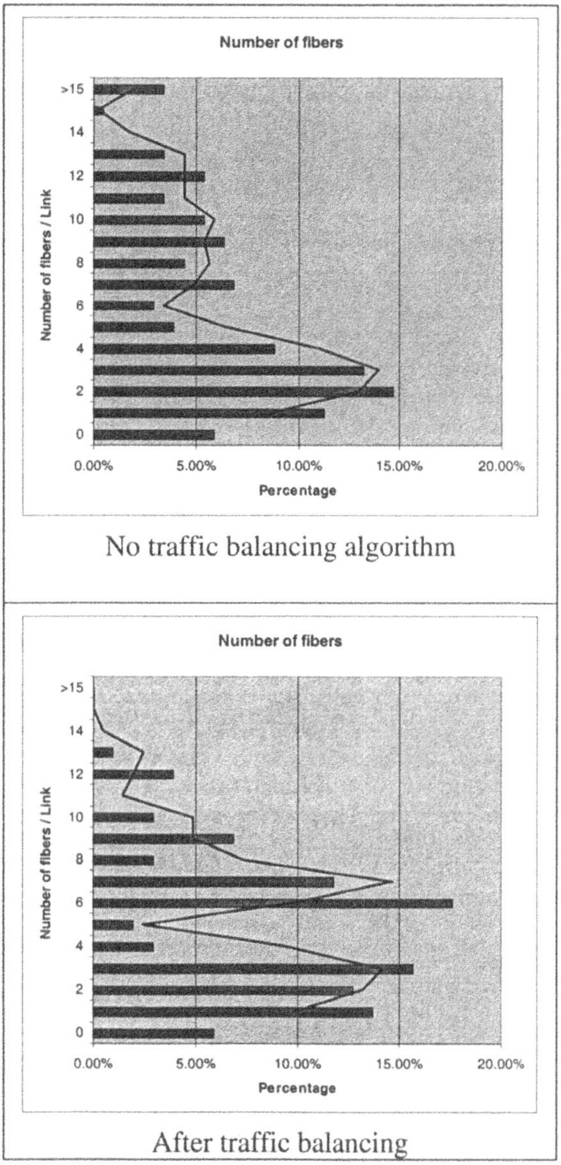

Figure 2. Ring dimensioning results

We have compared the results obtained by this algorithm with the LP formulation presented in the previous section. Although the two approaches find different solution to the TBWA problem, for every ring chosen to cover the network in Figure 1 the same number of fibers was found by both techniques. For these rings and given traffic demands, the proposed algorithm finds an optimum solution to the TBWA problem. To illustrate the difference between the results obtained by the LP formulation and the proposed algorithm, Figure 3, Tables 2 and 3 contain the routing and the wavelength assignment for one of the rings in the network. The objective of the ILP is not balancing the traffic in the ring, so in most of the cases routing and wavelength assignment obtained by each method will be different.

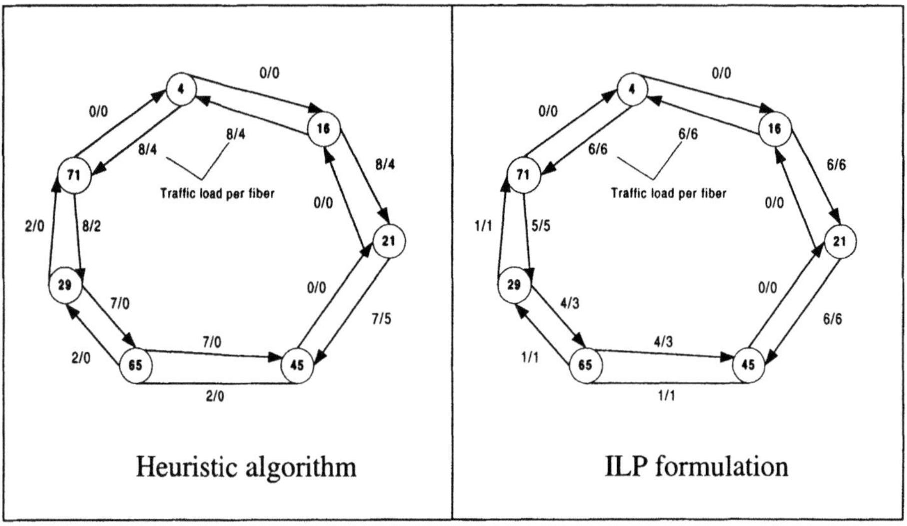

Figure 3. Ring 4: Links Loading

Traffic Demands	Fiber 1		Fiber 2	
	Clockwise direction	Counter-Clockwise direction	Clockwise direction	Counter-Clockwise direction
16->21: 1	w_1			
16->45:18	w_2, w_3, w_4, w_5, w_6, w_7, w_8	w_{10}, w_{11}, w_{12}, w_{13}, w_{14}, w_{15}, w_{16}	W_1, w_2, w_3, w_4	
16->29:3		w_9		w_9, w_{10},
16->71:2				w_{11}, w_{12}
21->45:1			w_5	
45->71:2	w_1, w_2,			
Protection in the ring	w_9, w_{10}, w_{11}, w_{12}, w_{13}, w_{14}, w_{15}, w_{16}	w_1, w_2, w_3, w_4, w_5, w_6, w_7, w_8	w_9, w_{10}, w_{11}, w_{12},	w_1, w_2, w_3, w_4, w_5

Table 2. Ring 4: Wavelength Assignment obtained by the heuristic algorithm

Traffic Demands	Fiber 1 Clockwise direction	Fiber 1 Counter-Clockwise direction	Fiber 2 Clockwise direction	Fiber 2 Counter-Clockwise direction
16->21: 1			w_1	
16->45:18	w_1, w_3, w_4, w_5, w_7, w_8	w_{11}, w_{12}, w_{14}, w_{16}	w_3, w_4, w_5, w_7, w_8	w_{12}, w_{14}, w_{16}
16->29:3		w_9		w_9, w_{15}
16->71:2		w_{15}		w_{10}
21->45:1			w_1	
45->71:2	w_4		w_1	
Protection in the ring	w_9, w_{11}, w_{12}, w_{14}, w_{15}, w_{16}	w_1, w_3, w_4, w_5, w_7, w_8	w_9, w_{10}, w_{12}, w_{14}, w_{15}, w_{16}	w_1, w_3, w_4, w_5, w_7, w_8

Table 3. Ring 4: Wavelength Assignment obtained by the ILP

4. CONCLUSIONS

Due to complexity and nature of the problem, the overall optimal design of an arbitrary topology based on WDM rings cannot be obtained for practical network sizes, in sub-second time. The paper has therefore proposed a basic framework for a fast, versatile and interactive ring network design tool that circumvents the complexity problem by subdividing the optimization problem into three sub-problems. Each sub-problem is optimally resolved individually and sequentially by means of relatively simple and fast algorithms. In particular, the paper has proposed a complete methodology for designing and dimensioning optical ring based networks. A practical example was shown based on the following assumptions. 1) The ring cover is selected with the objective to minimize the sum of the perimeters of the selected rings. 2) The routing of the traffic demands is obtained by minimizing the number of rings crossed by every demand. 3) The routing and wavelength assignment for the traffic demands within each ring is determined with the objective to minimize the number of fibers in the ring. Additional optimization modules may be added as and when the design objective changes. Moreover, with the proposed design methodology the intermediate results at the end of every optimization step can be observed and manipulated by the designer. In conclusion, the paper has presented a practical and versatile approach to designing WDM ring based networks.

5. REFERENCES

1. A. Fumagalli, I. Cerrutti, M. Tacca, F. Masetti, R. Jagannathan, S. Alagar. "Survivable Networks Based on Optimal Routing and WDM Self-Healing Rings", Proc. IEEE INFOCOM99.
2. A. Fumagalli, I. Cerrutti, M. Tacca. "Minimizing the number of Optical Crossconnect Ports in Mesh Networks Based on Bidirectional line-Switched WDM Self-Healing Ring Protection", Proc. IEEE ICCCN 99.
3. O. Gerstel, R. Ramaswami, G.H. Sasaki. "Fault Tolerant Multiwavelength Optical Rings with Limited Wavelength Conversion". IEEE JSAC., Vol. 16, No. 7, Sep 1998, pp. 1166-1178.
4. Hu-gon Kim; Dong-Wan Tcha. "Optimal balancing on SONET Bidirectional Rings". Operations Research, Vol 45, No.1, Jan-Feb 1997.
5. A. Lardies, A. Mata. "Optimization of SDH Networks With Unidirectional and Bidirectional Ring Structure", (Spanish) Proc. Telecom I+D 95.
6. L.M. Gardner, M. Heydari, J. Shah, I.H. Sudborough, I.G. Tollis, C. Xia. "Techniques for Finding Ring Covers in Survivable Networks", Proc. of IEEE Globecom 1994, pp. 1862-1866.
7. P.Mateti, N.Deo. "On Algorithms for Enumerating All Circuits of a Graph". SIAM J.Comput., Vol.5, No. 1, Mar 1976.
8. M. Gordran, M Minoux "Graphs and Algorithms". Wiley Interscience Publication, 1984, pp.12-21.
9. A.Fumagalli, L.Valcarenghi, "Fast Optimization of Survivable WDM Networks Based on Multiple Self-Healing Rings", Proc. SPIE, September 1999.
10. T.-H. Wu, "Fiber Optic Survivability", Artech House, 1992.

Dimensioning of Non-hierarchical Interconnected WDM Ring Networks

Peter Arijs, Piet Demeester
Department of Information Technology, University of Gent - IMEC
St-Pietersnieuwstraat 41, B-9000 Gent, Belgium
Tel: +32-9-267.35.91 - Fax: +32-9-267.35.99
E-mail: peter.arijs@intec.rug.ac.be

Philip Achten, Wim Van Parys
Comsof N.V.
Ketelvest 16, B-9000 Gent, Belgium
Tel: +32-9-234.13.56 Fax: +32-9-233.30.02

Key words: WDM rings, routing, dimensioning

Abstract: In this paper we consider the routing of wavelengths in an interconnected WDM ring network, based on Optical Channel Dedicated Protection Rings (OCh-DPRing). We developed an integer linear programming and an efficient heuristic solution method and compare both methods. Furthermore we present results on network cost and availability for different network and ring sizes.

1. INTRODUCTION

The Wavelength Division Multiplexing (WDM) technology enables to multiplex several optical signals together on one single fiber. Due to the increase in the amount of backbone traffic in recent years, public network operators have adopted WDM as an economic alternative to upgrade the capacity of the point-to-point fiber links in their long haul network [1]. As WDM technology further matures, all-optical networking will become possible through the use of optical add-drop multiplexers (OADM) and cross-connects (OXC), leading to a (re)configurable optical layer [2]. In such a configurable optical layer, data can be routed on a wavelength basis (along so-called lightpaths), and different client layer signals can be supported on separate wavelengths, such as SDH, PDH, ATM and IP traffic. Wavelength routing allows to lower the transmission costs by reducing the amount of optical-to-electrical conversions as well as the required amount of electrical processing capacity. As all-optical networking becomes a reality, also survivability in the optical layer will gain importance [2]. Due to the increase of traffic by the deployment of WDM, a single fiber cut potentially affects hundreds or thousands of client layer connections simultaneously. By using WDM recovery, all these failures can be tackled at once and no protection resources need to be allocated in the client layers. In addition, optical recovery also offers transparent survivable support for services which do not have their own recovery mechanisms.

A first step towards a survivable optical layer is seen through the use of WDM rings [3]. Since components and management for rings are simpler than for meshed networks, OADMs for ring architectures are expected to be earlier available at the market than OXCs. Indeed, fixed OADMs (capable of adding/dropping a limited set of fixed wavelengths) are already available from some vendors today, while reconfigurable OADMs - also capable of performing protection switching - are expected in the forthcoming years. First generation WDM ring architectures will be based on so-called optical channel dedicated protection rings (OCh-DPRing), protecting traffic on a wavelength basis. An OCh-DPRing consists of two counter-rotating fibers, with WDM deployed on each fiber. Each wavelength demand is protected using a main path along one side of the ring and a dedicated backup path along the other side of the ring. When a link or node failure occurs within the ring, the traffic can be switched over to the protection path, as illustrated in *Figure 1*.

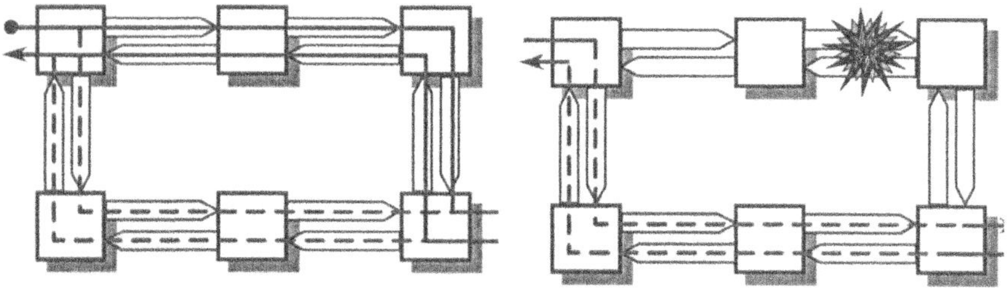

Figure 1a: OCh-DPring *Figure 1b*: OCh-DPring after failure

In the OCh-DPRing based on 1+1 protection, traffic is duplicated at the transmitter side (e.g. using an optical splitter) and the best of both signals is selected at the receiver side (e.g. using a 2x1 switch). Single ended protection switching occurs based on the monitoring information of the optical channel (hence no signaling protocol is required), which makes the implementation conceptually simple and allows fast protection switching.

The routing of bi-directional protected traffic on an OCh-DPRing is trivial since the main and protection paths of each demand, use both sides of the ring. As such, every demand is carried over every section of the optical ring. The utilization of every section - and the required ring capacity - is thus equal to the total demand. While single OCh-DPRings are straightforward to plan and operate, backbone networks based on OCh-DPRings typically consist of a multitude of interconnected rings. Building a network based on such a set of interconnected rings is a complex task, which consists of finding a minimum cost combination of interconnected rings, such that all demands in the network can be supported. While initially overlapping rings can be created such that no interconnection between rings is required, ring interconnection is a long-term requirement for an economical, reliable and manageable large-scale ring network. The rings can be interconnected logically in two ways, either by defining a hierarchy as in [4], or in a flat way without defining a hierarchy. In this paper we study the design of a flat interconnected ring network within an existing fiber topology.

The main tasks in the ring design process consist of finding communities of interest between which to deploy the rings (i.e. ring selection), and routing inter- and intra-ring traffic on the selected interconnected rings. In a hierarchical network, the second phase of this design process is trivial because the routing possibilities are constrained by the hierarchy. In this paper we consider this routing process for a flat ring network, which is a more difficult problem. The routing algorithms we developed can then be used as a phase in the ring selection process, or they can be used as a stand-alone process for dimensioning an existing network with already determined rings. When the routing algorithm is used as a phase in the ring selection process, it has to be executed a large number of times, in order to evaluate a large set of possible ring combinations. Therefore, it is of paramount importance that the routing algorithm is sufficiently fast. In this paper we present a fast and simple heuristic, that has shown to yield close to optimal results, and can thus be used for this purpose.

In section 2 of the paper, we discuss the givens and assumptions of the problem formulation. In section 3, solution methods are presented, both using integer linear programming and heuristic techniques. In section 4 we present some obtained results, by comparing the different algorithms and describing the effects of different network and ring sizes on network cost and availability. Finally, section 5 concludes the paper.

2. PROBLEM FORMULATION

We consider the physical network topology to be given as a set of cables (or links) L and a set of offices (or nodes) N. We also know the ring-sites R, i.e. the positions were OCh-DPRings can be deployed. If a site $r \in R$, passes through office $n \in N$, then $s(r,n) = 1$ (otherwise 0).

In a ring-site, an OCh-DPRing can be deployed by using one fiber pair in the cables, and one OADM in the offices along which it is routed. This can be seen in *Figure* 2, in which a simple 9-node network with 4 rings is shown. In the optical transmission section (OTS) layer, the physical network topology is represented, while in the optical channel (OCH) layer, the OCh-DPRings are represented.

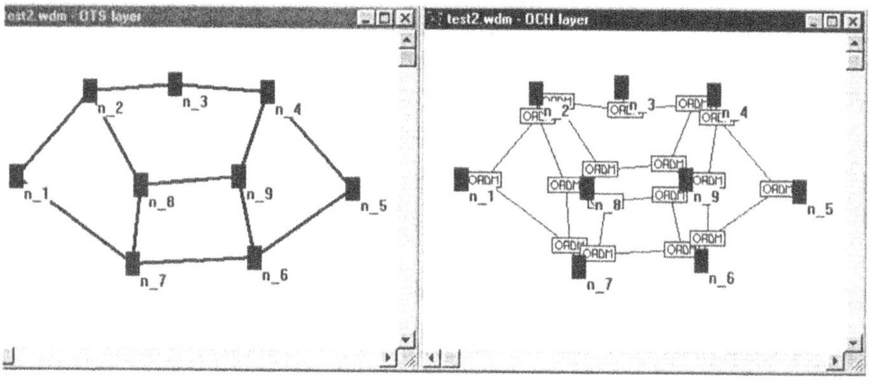

Figure 2: Physical network topology (OTS layer) and rings (OCH layer)

The demand d_{ij} expresses the amount of bi-directional wavelengths that have to be routed and protected between each two different offices $i \in N$ and $j \in N$ $(j>i)$. Each demand is protected in each ring in which it is routed (which allows to independently recover from multiple failures occurring in different rings) and traffic exchanged between 2 rings is protected using drop & continue as in SDH rings [5], which results in a very secure architecture.

We assume that at least one ring-site passes in each office, and the ring-sites are connected, such that all demands can be routed. The interconnection between two different rings is assumed to be hard-wired (i.e. without the use of cross-connects) and is performed in at least 2 different offices (to allow drop & continue).

We consider a network design based on a homogenous set of rings, i.e. all rings in the network are of the same type. In our case this means all rings are OCh-DPRings with the same amount of wavelengths W.

We consider the following generic cost model:

- A ring cost C_r^{ring} for each ring to be deployed on ring-site $r \in R$: this cost includes the fixed installation costs of the ring, fiber costs, amplifier costs, OADM costs, ...
- A routing cost C_r^{route} for routing one wavelength on a ring on ring-site $r \in R$: this costs accounts for the modular components of the ring that can be installed on a wavelength basis (e.g. tributary cards, transponders).
- An interconnection cost C^{inter} : the cost for interconnecting a wavelength between 2 rings (e.g. transponder cost).

The integer variable $x(r)$ represents the amount of OCh-PRings to be deployed in each ring-site $r \in R$. When multiple rings are deployed in the same ring-site r (i.e. when $x(r) > 1$), these rings are referred to as *stacked* rings.

Each demand d_{ij} can be routed along a number of possible paths $p_{ij} \in P_{ij}$. Such a path consists of a number of interconnected ring-sites (in section 3 we will discuss how these paths are obtained). If ring-site $r \in R$ is contained in path $p_{ij} \in P_{ij}$, then $q(p_{ij},r) = 1$ (otherwise 0). The integer variable $y(p_{ij})$ represents the amount of traffic that is routed on path $p_{ij} \in P_{ij}$.

The aim is to minimize the total cost of the interconnected ring network, such that all demands can be routed and protected. This can be formulated as follows:

$$\textbf{Minimize:} \begin{cases} \sum_{r\in R} C_r^{ring}.x(r) + \sum_{r\in R} C_r^{route}.\sum_{i\in N}\sum_{\substack{j\in N\\ j>i}}\sum_{p_{ij}\in P_{ij}} q(p_{ij},r).y(p_{ij}) \\ +2.C^{inter}.\sum_{i\in N}\sum_{\substack{j\in N\\ j>i}}\sum_{p_{ij}\in P_{ij}}\left(\sum_{r\in R} q(p_{ij},r)-1\right).y(p_{ij}) \end{cases}$$

Subject to:

$$\sum_{p_{ij}\in P_{ij}} y(p_{ij}) = d_{ij} \qquad \forall\, i\in N, j\in N \mid j>i$$

$$\sum_{i\in N}\sum_{\substack{j\in N\\ j>i}}\sum_{p_{ij}\in P_{ij}} q(p_{ij},r).y(p_{ij}) \le W.x(r) \qquad \forall\, r\in R$$

$y(p_{ij})$ integer and ≥ 0 $\qquad \forall\, i\in N, j\in N \mid j>i$

$x(r)$ integer and ≥ 0 $\qquad \forall\, r\in R$

The objective function consists of 3 terms. The first term accounts for the fixed ring cost contribution. The second term represents the cost of routing wavelengths on the rings. The third term adds the cost of interconnecting wavelengths between two rings. The factor 2 in this third term expresses the fact that interconnection between two rings takes place in two nodes (to allow drop & continue).

3. SOLUTION METHOD

3.1 Equivalent network

The *equivalent network* represents the topological relationship between the different ring-sites and can be used to assist in the routing decisions within the rings. In the equivalent network, each office is represented by a separate node. In addition, each ring-site is also represented by a node. Two types of links exist in the equivalent network. A link between a node representing an office n and a node representing a ring-site r is present if $s(r,n) = 1$. Such a link is called an *access link* and represents the fact that the ring-site passes through this office (and thus traffic injected at this office has access to the rings of the ring-site). A link between two nodes representing a ring-site is called an *interconnection link* and represents the fact that both ring-sites are interconnected in at least two offices (and thus traffic can be exchanged between rings of both ring-sites, using drop & continue). An example of a ring network and the corresponding equivalent network is given in *Figure 3*.

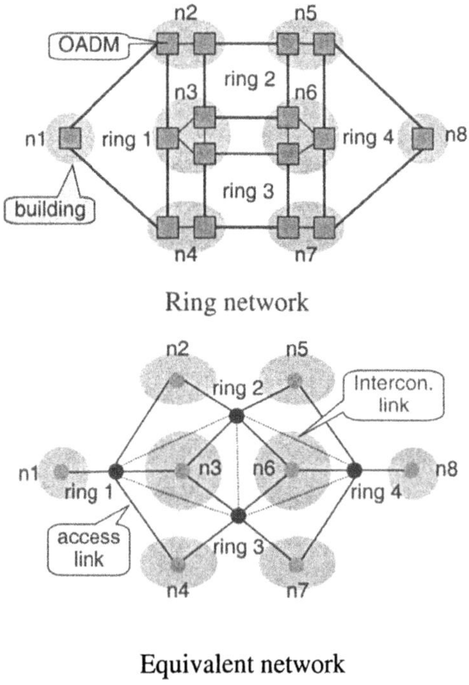

Figure 3: Ring network and equivalent network

Traffic between two offices in the ring network can be routed by finding a path between the corresponding nodes in the equivalent network. A path in the equivalent network should always start and end with an access link, while all intermediate links should be interconnection links. Afterwards, the obtained path in the equivalent network can be translated again to a sequence of interconnected rings in the original network.

We assign weights w_l and capacities c_l to links l in the equivalent network in the following way:

If l is an access link between nodes representing office n and ring-site r:

$$w_l = \frac{C_r^{ring}}{2.W} + \frac{C_r^{route}}{2}$$

$$c_l = W.x(r)$$

If l is an interconnection link between nodes representing ring-site r_1 and r_2:

$$w_l = \frac{C_{r1}^{ring}}{2.W} + \frac{C_{r1}^{route}}{2} + \frac{C_{r2}^{ring}}{2.W} + \frac{C_{r2}^{route}}{2} + 2.C^{inter}$$

$$c_l = W.\min\{x(r_1), x(r_2)\}$$

As such, the cost of a ring is divided over the links which are adjacent to the corresponding node in the equivalent network. The fixed cost is divided by the amount of wavelengths on the ring, distributing this cost equally among all channels on the ring. The interconnection costs are in turn accredited to the interconnection links.

3.2 Integer linear programming (ILP) solution

A first way of solving the problem under study is to implement the model presented in section 2 in an integer linear programming solver such as CPLEX [6]. The set of paths P_{ij} we use for each pair of offices is determined by searching for the K-shortest paths between the corresponding nodes in the equivalent network. The 'optimal' solution can only be found by considering all possible paths (i.e. K very large), however this tends to increase the computational complexity of the problem considerably. In practice, only modest values of K will be considered, which can still result in very good results, as will be shown. The ILP solution method is suitable for optimizing medium sized problems and to benchmark the heuristic solution method.

3.3 Heuristic solution method

For large-scale problem instances (large networks with a lot of ring-sites) the ILP solution method does not scale very well. For this purpose a fast heuristic solution method has been implemented. A schematic overview of the different phases of the heuristic can be found in *Figure 4*.

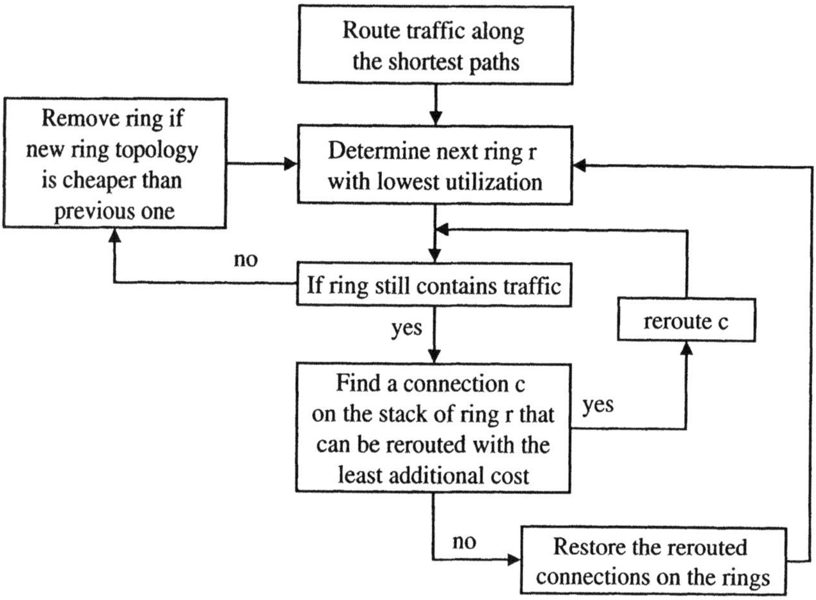

Figure 4: Heuri•tic optimization method

The heuristic starts by routing the connections d_{ij} along their shortest path in the equivalent network. After all connections have been routed we can determine the amount of traffic routed on each ring-site. From this we can deduct the amount of stacked rings required at each ring-site, as well as the utilization of each ring. In case multiple stacked rings are present at a ring-site, all rings have a utilization of 100%, expect for the last ring in the stack, which might have a lower utilization. The heuristic now attempts to eliminate some of the rings with low utilization ratios, by trying to reroute connections on this ring (or on any other ring of the stack) on a different route, not containing this ring-site. On the ring with the *lowest utilization* ratio, we start with the connection that can be rerouted on an alternative route, with the *least additional cost* compared to the cost of its original route. The new route is calculated in the equivalent network (along the shortest path), in which we first tear down the original route (and adapt the capacities), and from which we exclude the current ring and all other rings that have no free capacity left. This process of rerouting connections is repeated until no more traffic is present on the last ring of the stack, or until no more connections can be rerouted. In the first case the ring is removed if the elimination of the ring leads to a cheaper overall solution. If the removal does not lead to a cheaper solution, or in the second case, we restore the rerouted connections on the current ring and do not remove the ring. In both cases the algorithm is continued by searching for the ring with the next lowest utilization, and again we try to reroute connections on this ring in the same way. This process is repeated until all rings are evaluated. Note that the utilization ratios are recalculated in each iteration, because the rerouting of traffic of removed rings also affects the traffic on the other rings.

4. RESULTS

In this section, we present results for 2 sample networks:

- a network with 16 nodes, 21 links and a traffic matrix containing 116 wavelengths
- a network with 32 nodes, 41 links and a traffic matrix containing 256 wavelengths

We consider 16 wavelength OCh-DPRings to be used throughout the network. For the 16-node network, we considered a design based on 3 large ring-sites or on 7 smaller ring-sites. For the 32-node network, we considered a design based on 7 large ring-sites or on 13 smaller ring-sites.

We used a generic cost model, with the cost of a ring r passing through N_r offices and having A_r amplification sites:

$$C_r^{ring} = N_r.(C^{link} + C^{OADM}) + A_r.C^{AMP} \quad \text{and} \quad C_r^{route} = N_r.C^{route}$$

The used cost values for this case study are: $C^{link} = 200$, $C^{OADM} = 50$, $C^{AMP} = 40$, $C^{route} = 20$ and $C^{inter} = 10$.

In *Figure 5*, we present a cost comparison of the different designs, using both the heuristic and the ILP algorithm based on the K-shortest path algorithm (using K=2 and K=4). The graph also shows the utilization ratios[1] of the different networks. A first observation is that the heuristic yields results, which are very close or equal to the results yielded by the ILP algorithm. While the heuristic typically takes a couple of seconds or less on a 166 MHz Pentium PC, the ILP algorithm takes some more time, but never more than a couple of minutes for the considered networks. A second observation is that the ILP algorithm with K = 2 (thus only considering the shortest and second shortest path in the equivalent network) gives in most

[1] Utilization = total used capacity / total installed capacity

cases the same result than for higher values of K (although higher values of K require longer calculation times).

When comparing the network designs based on different ring sets, we can conclude that a network composed of a larger amount of relatively small rings results in a lower overall cost, than a network composed of a small number of larger rings. Although in a network with larger rings, less inter-ring traffic is present, and thus a lot of traffic can be carried on a single ring, the total cost is higher. This is because connections typically have long protection paths on the rings, thus making inefficient use of the network capacity. Even in a single ring, a connection between neighbouring nodes takes up capacity on the entire long ring. In addition the connections that do require to be routed on multiple rings, consume capacity on almost the entire network. In a network with smaller rings, the protection paths on each ring are relatively short, and even if a connection is transported on multiple rings, the total amount of protection capacity remains reasonable. However, it could be noted that one or two large 'overlay' rings might be beneficial for transporting connections from one end of the network towards the other end in a network composed of smaller rings.

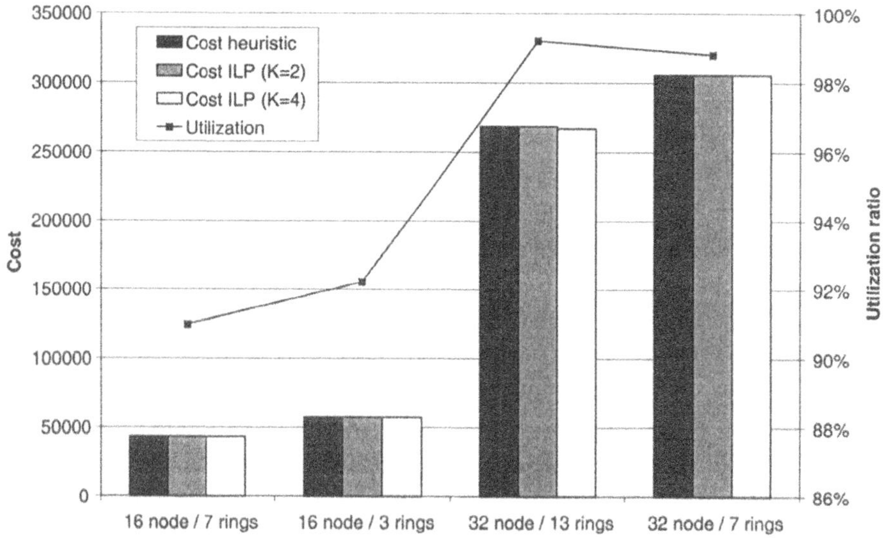

Figure 5: Cost comparison

Smaller rings do not only result in a lower overall network cost, they also help in increasing the availability of the network. This is reflected in *Figure 6*, which represents the expected loss of traffic (ELT), which is the amount of traffic that the network is expected to lose per year due to failures [7]. We express ELT in STM-16 hours per year, assuming each wavelength transports one STM-16 signal (=2.5 Gb/s). We have calculated the ELT both for link failures only, and for link and node failures, while considering ring interconnection with and without drop & continue. The availability of a network element is reflected by its mean time between failures (MTBF) and mean time to repair (MTTR):

- Links: MTBF: 1 failure every 300 km per year; MTTR: 24 hours
- OADM: MTBF: 100000 hours; MTTR: 6 hours
- Optical amplifier: 500000 hours; MTTR: 24 hours

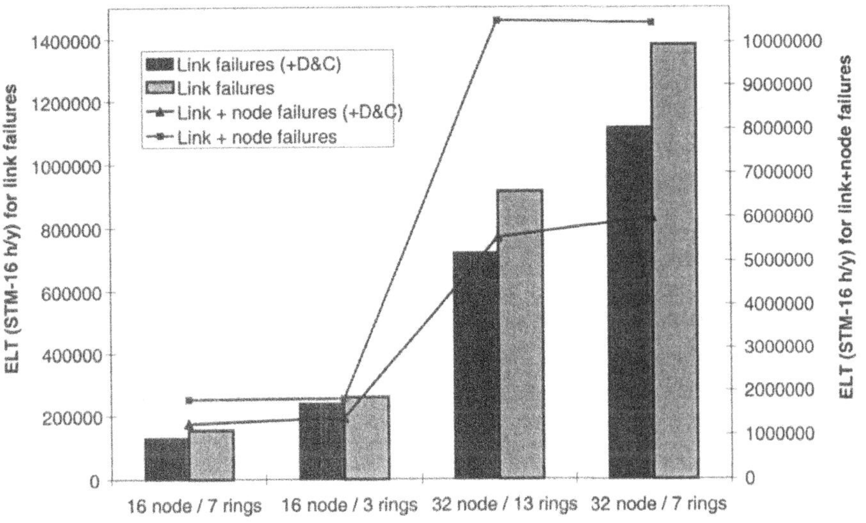

Figure 6: Availability comparison

From *Figure 6* we can indeed conclude that a network composed of a large amount of small rings results in a lower ELT, and thus a higher network availability. This is particularly evident when only considering link failures, as multiple small rings can better survive from multiple link failures in the network, because the chance is higher that the different link failures occur in different rings. When considering both link and node failures, the effect is less strong because a node failure involves the loss of all traffic originating/terminating in that node, which is traffic that can not be recovered any way. This explains the much higher ELT values for the node failures and the small differences between the networks with different ring sizes. In fact the difference in ELT for link and node failures between large rings and small rings is very small when not considering drop & continue. Indeed, larger rings can survive from less multiple failures, but on the other hand, they involve less ring interconnections (single point of failure in case no drop & continue is used). When drop & continue is used the difference between small and large rings is more apparent. The overall effect of drop & continue on the ELT values is remarkable, and increases for the larger network. When considering both link and node failures, the ELT almost doubles when not using drop & continue. As drop & continue does not require any additional ring capacity for OCh-DPRings, it can thus provide a substantial increase in availability at a relatively low cost.

5. CONCLUSION

We discussed the routing problem in interconnected OCh-DPRing networks. We have developed an ILP solution method based on the K shortest paths, that gives good results for low values of K. Furthermore, a simple and fast heuristic optimization method was developed that performs in close range with the optimal result yielded by the ILP algorithm. The heuristic is sufficiently fast to be used as a phase in the more complicated ring identification process, during which it has to be executed a large number of times.

When comparing routing in a network with a limited set of large rings or a large set of smaller rings, the latter architecture is preferred both from a cost and availability point of view. Also the impact of drop & continue on the overall network availability is substantial and increases as the size of the network grows.

ACKNOWLEDGMENT

Bart Van Caenegem (University of Gent), Eugenio Iannone and Filippo Bentivoglio (Pirelli) are acknowledged for the fruitful collaboration.

REFERENCES

[1] E. Lowe, "Current European WDM deployment trends", *IEEE Communications Magazine*, vol. 36, no. 2, Feb. 1998, pp. 46-50.

[2] P. Demeester, T.-H. Wu, N.Yoshikai (eds.), "Survivable communication networks", *Special issue of IEEE Communications Magazine*, Vol. 37, No. 8, Aug. 1999.

[3] P. Bonenfant, "Protection and restoration in optical networks", *Proc. of OFC'99* (San Diego, CA, Feb. 1999), Tutorial Sessions, pp. 199-214.

[4] B. Van Caenegem, P. Demeester, "Design of interconnected WDM ring networks", *Proc. of ONDM'99* (Paris, France, Feb. 1999), pp.61-68.

[5] M. Sexton and A. Reid, "Broadband networking: ATM, SDH and SONET", *Artech House*, London, ISBN: 0890065780.

[6] ILOG Inc., CPLEX Division, "Using the CPLEX callable library", 1997.

[7] D. Vercauteren et al. "Evaluation environment for telecommunications networks", *Proc. of Networks'96* (Sydney, Australia, Nov. 1996), pp. 671-676.

A Functional Model for the SONATA Switchless Optical Network

S. De Maesschalck, P. Demeester
Dep. of Information Technology, University of Gent – IMEC
Sint-Pietersnieuwstraat 41, B-9000 Gent, Belgium
Tel: +32 9 267 35 90, Fax: +32 9 267 35 99
E-mail: sophie.demaesschalck@intec.rug.ac.be

Key words: Switchless optical network, functional model, management advantages

Abstract: In this paper we clarify the concept of the switchless optical network developed in the framework of the ACTS project SONATA. The functional model we developed for this kind of network is explained. Some management advantages of this kind of network structure are discussed.

1. INTRODUCTION

The growing need for bandwidth has led to the introduction of WDM (Wavelength Division Multiplexing) in the core network. This technique is for the moment mainly used in point-to-point links but in the future the introduction of optical cross-connects and add/drop multiplexers, replacing the current electrical nodes, may be expected. The general structure of the network however will not change: there will still be a distinction between access and core part of the network and between transport and switching function. The core network will consist of multiple large and complex optical nodes connected by high-capacity optical links in a mesh or ring structure. Several switching technologies (e.g. ATM, IP,...) will work in parallel on this transport layer. The access network will be based on different solutions like ATM PONs, xDSL on copper pairs,....

The switchless optical network proposed here, is a new and simplified network structure developed in the framework of the ACTS project SONATA to avoid the need for large and fast switching nodes in the network [1], [2]. The distinction between access and transport part of the network has completely disappeared. It is basically a single layer all-optical network providing end-to-end optical connections between a large number of terminals distributed over a wide geographical area. This multiple-access network employs both time and wavelength agility at the terminals (a TDMA/WDMA protocol) to permit customers to send both connection-oriented and datagram traffic over the network.

The network has a very simple structure with all traffic passing through a single node providing passive wavelength routing and, optionally, actively controlled wavelength conversion. The network is switchless in the sense that there is no need for electrical or optical switches or cross-connects in the network, except for the single central routing node.

The aim of this paper is to provide a functional model for this type of network based on the modelling methodology described in ITU-T Recommendation G.805 [3]. The use of this type of network modelling is very powerful as a foundation for developing network management specifications [4]. In section 2 the structure and the concept of SONATA are further explained. In section 3 the layer architecture is discussed. Section 4 describes the functional model developed for this network. In section 5 some management advantages of the switchless optical network structure are discussed.

2. STRUCTURE AND CONCEPT

The structure of a basic switchless optical network is shown in *Figure 1*. On the figure the network structure has been unfolded in a downstream and an upstream part to enhance its legibility. The direction of all signals is from left to right.

The heart of the switchless optical network is the passive wavelength routing node (PWRN) to which a number (N_d) of PONs (Passive Optical Network) are connected. Each PON has a separate fibre for the upstream and downstream direction. The customer's terminals, each containing a wavelength agile sender and transmitter, are attached to the branches of those PONs. Each PON can transfer data to each other PON (including his own) on a dedicated wavelength that traverses the central node. As the terminals of each PON have to be able to communicate with the terminals of all PONs, a terminal needs a tuneable sender/receiver pair able to transmit/receive on N_d different wavelengths. On top of this, each wavelength between a PON pair is shared by the various terminals of that PON which results in the need for a TDMA protocol. Time is divided into fixed-length timeslots and in a single timeslot the terminals of a PON need to use a different wavelength. A sequence of timeslots is further structured into fixed-size time frames on each wavelength channel.

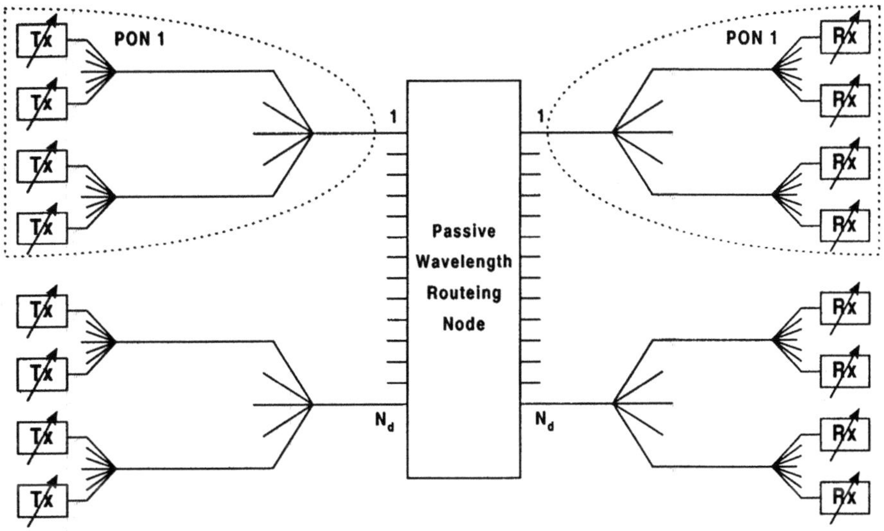

Figure 1. Basic structure of the switchless optical network

The different optical channels in a PON are multiplexed by the successive combiners into a single WDM signal per PON. The network thus uses a double multiplexing strategy: one in the time domain and one in the wavelength domain. This WDM signal enters the router at an in-going port, is demultiplexed and the optical channels are routed to the correct outgoing port, based solely on their wavelength, where they are multiplexed with the channels routed to the same outgoing port. This routing principle is illustrated in *Figure 2* for the switchless network structure of *Figure 1*. The WDM signal leaving the central node at each outgoing port is broadcast to all terminals of the PON to which the destination terminal belongs.

		N_d outgoing ports				
		1'	2'		N_{d-1}'	N_d'
N_d in-going ports	1	λ_1	λ_2		λ_{Nd-1}	λ_{Nd}
	2	λ_{Nd}	λ_1		λ_{Nd-2}	λ_{Nd-1}
	N_{d-1}'	λ_3	λ_4		λ_1	λ_2
	N_d	λ_2	λ_3		λ_{Nd}	λ_1

Figure 2. Fixed routing of the wavelength channels in the central node for the basic structure

As illustrated in *Figure 3* the SONATA switchless optical network has a number of additional features, described below.

In general cascaded optical amplifiers are required in the branches of the PONs due to the large geographical area covered and the large splitting ratio. In such large split "hyper-PONs" a noise gating functional block has to be placed after those amplifiers and before the last combiner in the upstream direction to avoid accumulation of ASE noise from the not-used branches during each timeslot [5].

The central node also needs a connection to the central network controller, which is responsible for the allocations of the time slots in such a way that the network is collision free.

In order to increase the flexibility and the available bandwidth between a PON pair, additional wavelength channels (N_i) are introduced that are dynamically allocated to a PON pair depending on the traffic requirements. The multiplex coming from a PON now contains N_d+N_i optical channels. The central node therefore needs N_i extra in-going and outgoing ports, called dummy ports, each of which is connected to an array of wavelength converters. The complete structure of the SONATA switchless optical network is shown in *Figure 3*.

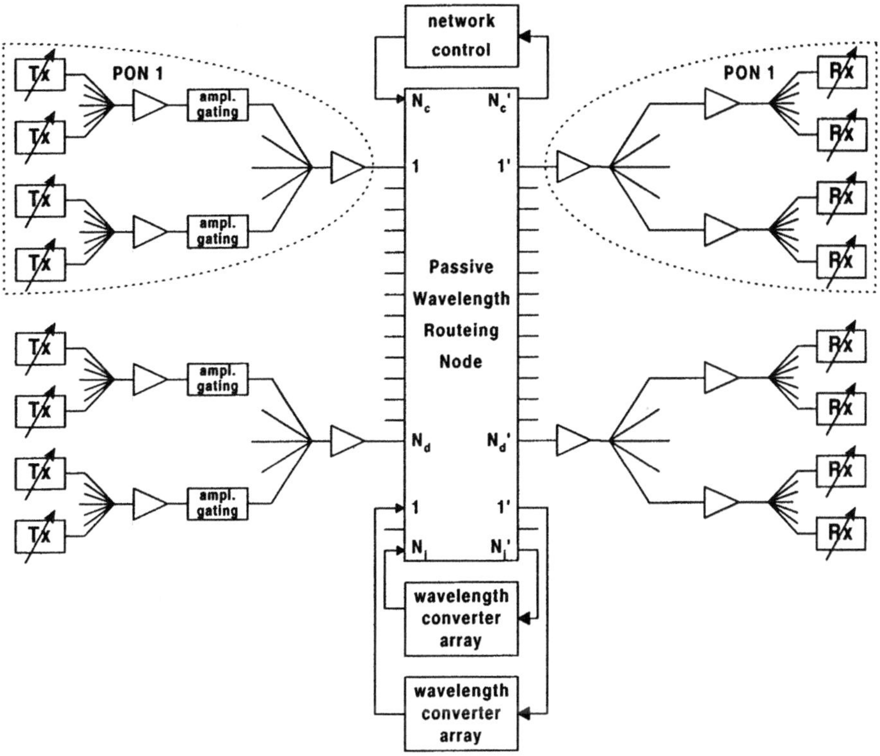

Figure 3. Complete structure of the switchless optical network

The extra optical channels are routed across the PWRN to the correct outgoing dummy port, based on their wavelength and pass through the actively controlled wavelength converter block. Here their wavelength is translated to the wavelength needed to reach the correct outgoing port of the PWRN. The routing table of *Figure 2* is extended in *Figure 4* for the complete structure.

To differentiate them from the (direct) wavelength channels that only pass the PWRN once, these wavelength channels are called indirect wavelength channels, because they also pass trough the wavelength converter bank.

		N_d outgoing ports					N_i outgoing dummy ports		
		1'	2'		N_{d-1}'	N_d'	1'		N_i'
N_d in-going ports	1	λ_1	λ_2		λ_{Nd-1}	λ_{Nd}	λ_{Nd+1}		λ_{Nd+Ni}
	2	λ_{Nd+Ni}	λ_1		λ_{Nd-2}	λ_{Nd-1}	λ_{Nd}		$\lambda_{Nd+Ni-1}$
	N_{d-1}'	λ_{Ni+3}	λ_{Ni+4}		λ_1	λ_2	λ_3		λ_{Ni+2}
	N_d	λ_{Ni+2}	λ_{Ni+3}		λ_{Nd+Ni}	λ_1	λ_2		λ_{Ni+1}
N_i in-going dummy ports	1	λ_{Ni+1}	λ_{Ni+2}		$\lambda_{Nd+Ni-1}$	λ_{Nd+Ni}	λ_1		λ_{Ni}
	N_i	λ_2	λ_3		λ_{Nd}	λ_{Nd+1}	λ_{Nd+2}		λ_1

Figure 4. Fixed routing of the wavelength channels in the central node for the complete structure

With one additional wavelength channel (Ni=1) it is e.g. possible to double the available bandwidth for in-PON traffic. We will explain with an example how this works. Traffic within PON 1 is transmitted on the direct wavelength λ_1. The PWRN routes this optical channel from port 1 to port 1'.With one additional wavelength available data from PON 1 is sent on λ_{Nd+1}. It enters the PWRN on port 1, like all traffic coming from PON1, and is routed to outgoing dummy port 1' (see figure 4). In order to be transmitted to PON1, it has to reach outgoing port 1'. Therefore it has to re-enter the PWRN with wavelength $\lambda_{Ni+1}=\lambda_2$. Thus, the wavelength converter has to translate λ_{Nd+1} to λ_2.

To further increase the flexibility and not to allocate an entire wavelength when only a fraction of its bandwidth is needed, the indirect wavelengths will be shared among different PON pairs. In this way one extra optical channel can e.g. provide extra capacity between PONa and PONb in the first half of the frame and between PONa and PONc in the second half.

A national scale switchless optical network could have following dimensions:

- $N_d = 400$ ports of the central node are connected to a PON
- Each PON is connected to 50 000 terminals
- $N_c = 1$ port of the PWRN is connected to the network controller
- $N_i = 400$ ports of the central node are connected to wavelength converter blocks

- The central node is a 801x801 routing node with 801 wavelength channels on each in-going and outgoing port and 0,5 nm spacing
- The maximum distance between two terminals is about 1000 km.
- Burst mode operation at 622 Mbit/s on an optical channel.
- 10 Mbit/s total average rate per terminal
- 622 Mbit/s maximum rate per terminal
- 200Tbit/s maximum network throughput

3. LAYER ARCHITECTURE

A first important task is to find a suitable layering architecture for the switchless optical network. As mentioned before this network is considered a single-layer network, so it seems contradictory to talk about layering. Perhaps the term single-technology network would be better suited than single-layer network as the data transferred over the network stays in optical form from source to destination terminal. But within this optical transmission format several layers can be distinguished with a distinct functionality.

The SONATA concept is based on a mixture of WDM and TDM techniques to allow interconnection between the terminals of the various PONs. It seems a logical choice to provide separate layer networks for both the wavelength and the time dimension. The layering architecture for the switchless network consists then of the Optical Transport Network layers [6] (OTS, OMS and OCh layers) with a TDM layer on top of them.

4. FUNCTIONAL MODEL

Based on the layering architecture we developed a functional model for the components of a switchless optical network. *Figure 5* shows this functional model for the network of *Figure 3.* The model will be explained for the different elements encountered in the network.

Terminal transmitter

At the terminal transmitter, the client data format is adapted to the timeslot format used in the switchless optical network and stored in the terminal buffer. Based on the allocation scheme received from the central network controller, the wavelength agile transmitter sends each timeslot of information in a particular timeslot of the frame on a particular wavelength. The task of the transmitter is thus threefold and spread out over all layers: adaptation of the data to timeslots at the TDM/client adaptation function, selection of the right timeslot in the OCh/TDM adaptation function, selection of the correct wavelength on which to transmit at the OMS/OCh adaptation function.

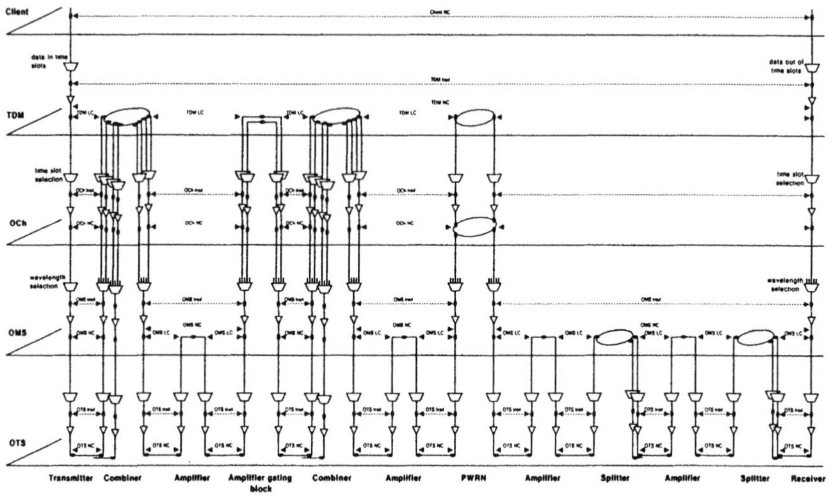

Figure 5. Functional model for the switchless optical network

Combiner

At the combiner, the signals coming from several terminals within a PON are combined. This component acts as a TDM switch whereby the timeslots on the same wavelength but in a different optical multiplex section are routed to a single transit group that is associated with a single adaptation function which time multiplexes the timeslots into the outgoing optical channel [4]. Those optical channels are then multiplexed together in on optical multiplex. This is illustrated in *Figure 6.*

Amplifier

The next network element encountered by the signal transmitted is the optical amplifier. Since the WDM signal is amplified as a whole, the functionality of this component has to be represented at the OMS level.

Amplifier gating block

At the amplifier gating block, the optical signal is demultiplexed. This amplifier gating block is responsible for gating the optical burst signal so that the transmission path is interrupted when no signal is present. Since gating is done timeslot by timeslot, the signal needs to be adapted out of the lower layers into the TDM layer.

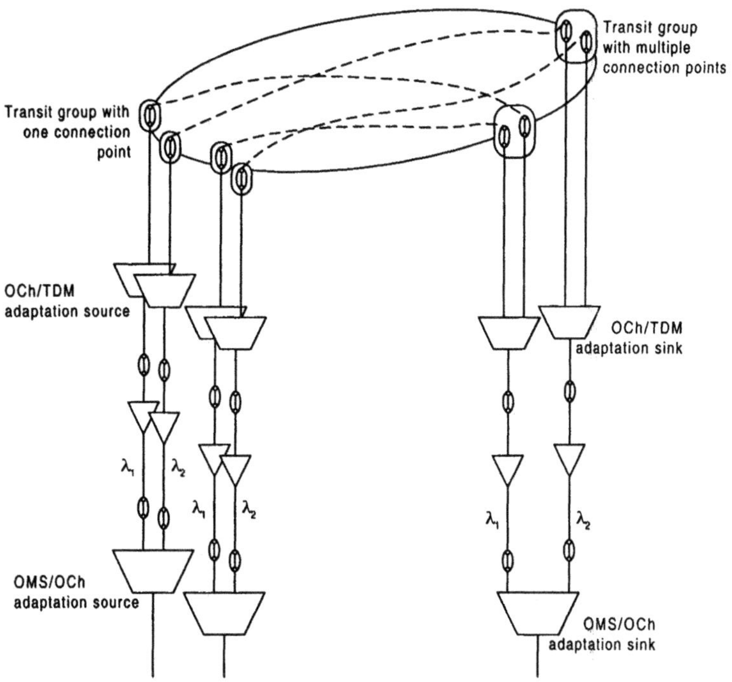

Figure 6. A combiner

PWRN

The signal then passes the central passive wavelength routing node. A mentioned before, a timeslot can be transmitted on a direct or on an indirect wavelength. In the former case the timeslot passes trough the central node only once and its wavelength cannot be changed. In the latter case the timeslot runs trough a wavelength converter. Besides this, an indirect wavelength channel can be dynamically reconfigured. This means that during one timeslot the wavelength interconnects PONx and PONy, while one timeslot later it provides interconnection between PONx and PONz. This difference has as consequence that on a direct wavelength routing is solely based on the wavelength, which has to be represented at the OCh level, while on an indirect wavelength routing is performed on a timeslot basis, which has to be represented at the TDM level.

The functional model of the PWRN is illustrated in *Figure 7* for a switchless network with two PONs and one additional wavelength.

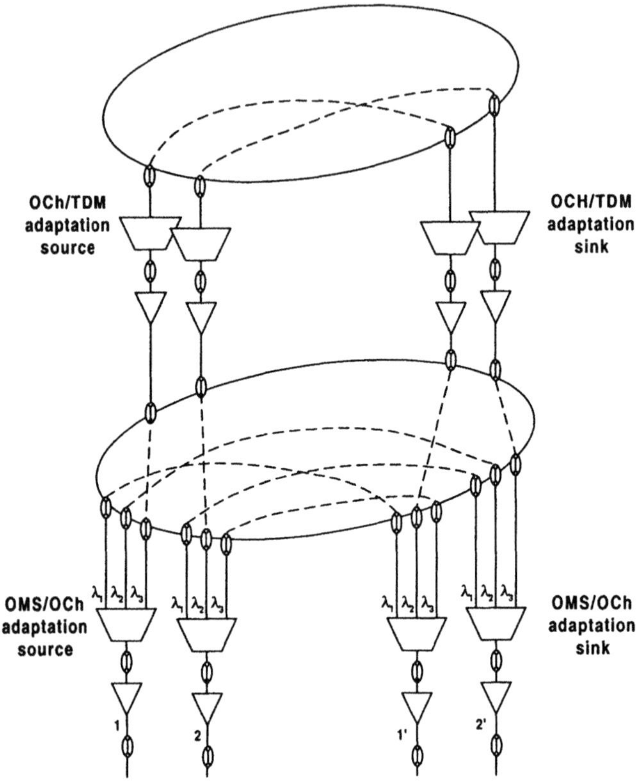

Figure 7. The PWRN

We take a look at what happens with the traffic coming from PON1. The multiplex section at in-going port 1 is demultiplexed into three optical channels. λ_1 is routed directly to outgoing port 1' and λ_2 is routed directly to outgoing port 2' at the OCH layer. The routing in the matrix at this layer is fixed. λ_3 is the indirect wavelength. Some timeslots per frame on this wavelength must be routed to outgoing port 1', some to outgoing port 2'. Therefore the optical channel has to be adapted to the TDM layer, where the individual timeslots are routed correctly. The routing in the matrix at the TDM layer is not fixed but will change over time.

Splitter

In the downstream direction the first new network element is the splitter. The signal at the entrance of the splitter is a multiplex of $N = N_d + N_i$ wavelengths. The traffic entity that is broadcast to each of the branches of the splitter is the characteristic information of the optical multiplex section.

The splitter therefore provides a broadcast function at this level via an OMS subnetwork [4].

Terminal receiver

Finally the time slotted signal reaches the destination terminal. At the receiver, the right wavelength is chosen from the optical multiplex (based again on an allocation scheme coming from the central network controller) at the OMS/OCh adaptation function. The right timeslot in the frame is selected at the OCh/TDM adaptation function. The information in the timeslot is finally adapted out of the timeslot format back into the client layer transmission format.

5. MANAGEMENT ADVANTAGES

The switchless optical network is basically a single-technology all-optical network. Because of this, it has some management advantages not found in other network structures.

The shared access concept usually found in access networks is here extended to the entire network. All resources of the network are shared which results in a lower cost compared to more traditional networks.

Data, in the format of fixed length timeslots, is transported over the network from transmitting to receiving terminal without buffering. The timeslots to be transmitted are buffered at the edge of the network and once a timeslot is taken out of this buffer it is transmitted without any further form of buffering across the entire network until it reaches the terminal buffer at the other end of the network.

The timeslot format used can transport all kinds of formats and protocols, both connection-oriented and connectionless. In the case of connection-oriented traffic, an integer number of slots per frame is requested at connection set-up. At termination of the connection a deallocation request has to be sent to release these time-slots. For datagram traffic, a number of slots is allocated for the duration of one frame only.

Because of the reduced number of architecture layers in the network, there are no problems due to interworking and co-ordination of the management system of different technologies.

6. CONCLUSION

The switchless optical network represents an alternative approach to networking meant to drastically simplify the structure of future broadband networks. It is a single layer network in which all switching functionality is removed from the network nodes and distributed at the terminals, ending in a network structure with a single node.

The paper presented, for the first time, a functional model for this network in a four-layer architecture. The model for the various network components was discussed. Some management advantages of this network structure were discussed.

7. ACKNOWLEDGEMENTS

This work has been supported by the European Union ACTS Program through the project SONATA. The authors warmly acknowledge all their colleagues in the SONATA project.

8. REFERENCES

[1] A.M. Hill, S. Carter, J. Armitage, M.Shabeer, R.A. Harmon and P. Roise, "A scalable and switchless optical network structure employing a single 32x32 free-space grating multiplexer", IEEE Photonics Technology Letters, Vol.8, No.4, P 596-572, 1996

[2] N.P. Caponio, A. M. Hill, R. Sabella, "Switchless optical network for advanced transport architecture, IEEE/LEOS'98, Orlando, FL, USA, December 1998

[3] ITU-T Rec. G.805, "Generic Functional Architecture of Transport Networks", November 1995

[4] Private communication with A. McGuire, BT, September 1999

[5] A.M. Hill, "WDM networking in the access network and beyond", ECOC'95 Optical Networking Forum Workshop, Brussels, 22nd September 1995, paper S1.4

[6] ITU-T Draft Rec. G.872, "Architecture of Optical Transport Networks", February 1999

Using Normalised Sections for the Design of All Optical Networks

C. Caspar, R. Freund*, N. Hanik**, L. Molle*, C. Peucheret***

Heinrich-Hertz-Institut für Nachrichtentechnik GmbH, Berlin; e-mail: caspar@hhi.de
**Virtual Photonics Incorporated, Berlin; e-mail: r.freund@virtualphotonics.com*
*** T-Nova, Technologiezentrum, Berlin; e-mail: norbert.hanik@telekom.de*
**** Technical University of Denmark, Copenhagen; e-mail: cp@com.dtu.dk*

Key words: optical networks, dispersion compensation, NRZ, numerical optimisation

Abstract: A novel concept for transparent link design is presented, and evaluated numerically and experimentally. 10 Gbit/s single channel transmission over more than 4000 km of Standard Single Mode Fibre is demonstrated. At reduced transmission distances, the systems show a high robustness against variations of the system parameters.

1. INTRODUCTION

Transparent optical WDM-systems are extremely increasing capacity and flexibility of today's information networks. The drawback of the network evolution towards transparent WDM is, however, that the almost perfect degree of network stability of periodically regenerated SDH-networks is lost. Concerning network planning, the complex interaction of several physical effects leads to specific limitations of the transmission distance to the so-called 'transparency-length'. In order to make wide use of the benefits of transparency, but also to avoid network instabilities, a future WDM-Network will consist of Transparent Optical Domains (TD) without 3R-regeneration, where a wide variety of data rates, data formats and wavelengths can be transported, and Subnetwork Interfaces (SNI) with 3R-regeneration for the interconnection of these Transparent Domains (*Figure 1*).

A strategy to simplify transparent network design and operation is to define so called 'Normalised Network Sections' which are cascaded to compose a particular transparent domain [1].

The optimisation of a specific Normalised Section is done in 2 steps: *First, numerical simulations* are performed for a *single channel* transmission system to identify the optimum parameter range and the system tolerances that can be expected. Due to this restriction, the calculation time is still within tolerable limits. In a *second* step, these numerical results are checked by *experiments*. Here, recirculating loop experiments were realised to imitate an arbitrary number of cascaded Normalised Transmission Sections. Finally, the performance of a WDM system using the so designed normalised section will show the limits for WDM transmission. Here, additional effects, e. g. Raman Scattering and Cross Phase Modulation [2], have to be taken into account. Nevertheless, the optimum range of operation parameters does not change significantly. In this paper, the optimisation of Single Channel Transmission is demonstrated.

The final result of this strategy is profound knowledge about the maximum extension of an optical network which applies the respective Normalised Transmission Section, and the parameter tolerances for network-operation. The next section shows an example for this promising approach.

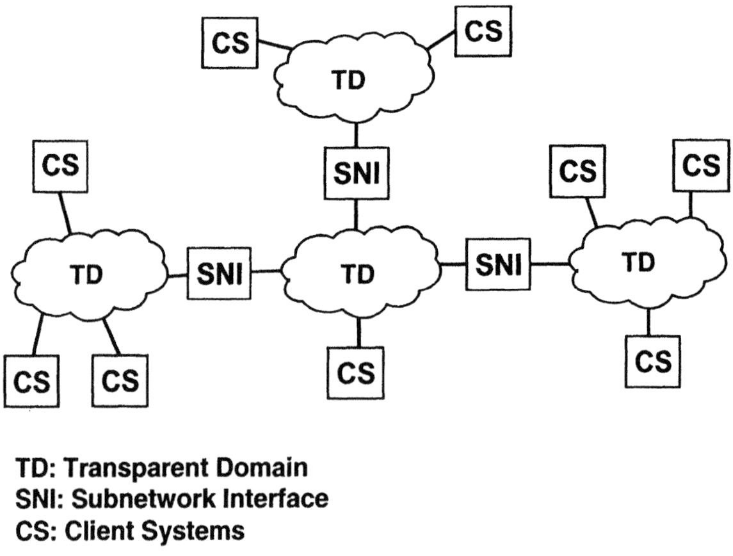

Figure 1. Vision of a future Optical Network

2. DESIGN AND OPTIMISATION OF THE TRANSPARENT DOMAIN

2.1 Specification of Normalised Transmission Sections

As the design is performed for a maximum channel data rate of 10 Gbit/s, the normalised sections have to be equipped with dispersion compensating modules. The arrangement, however, of deployed Standard Single Mode fibres (SMF, which are used in most of the European network), chromatic Dispersion Compensating Fibres (DCF) and Erbium Doped Fibre Amplifiers (EDFA) to compensate for the fibre attenuation, is variable, and may be a compromise between performance and costs.

Figure 2 shows two examples of normalised transmission sections based on SMF and Dispersion Compensating Fibre (DCF) using different schemes for the chromatic dispersion compensation. In both cases, the compensation ratio (CR) as well as the SMF/DCF input power can be varied independently. In case of pre-compensation, the additional (dotted) SMF in the transmitter section and the corresponding DCF in the receiver section allow changes of the so called "pre-distortion".

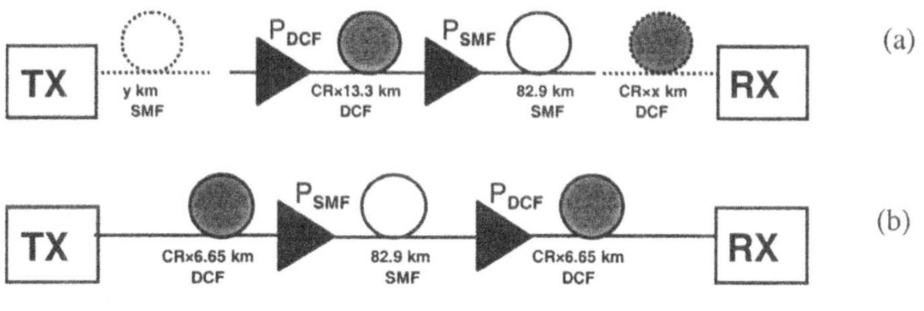

Figure 2. Normalised sections using pre-compensation (a) and hybrid-split (pre- and post-) compensation technique (b)

2.2 Numerical investigation of the Normalised Section

Objective of the optimisation process is to find the maximum number of cascadable sections with sufficient signal quality at the receiver. Therefore the power launched into the SMF, the power launched into the DCF, and the dispersion compensation ratio were independently varied. *Table 1* gives an overview of the system parameters of the experimental setup (section 2.3) which were also used in the simulations. This numerical simulations were done by solving the Nonlinear Schroedinger Equation [3] applying the Split-Step-Fourier method. Due to the high SMF/DCF dispersion coefficient, the simulations are based on the assumption of independent signal and noise propagation. The Q-factor [4] was calculated under the assumption of Gaussian noise at the receiver, which gives a rather conservative estimation of the system performance.

Table 1. Parameters of the loop set-up (s. *Figure 5*), used in simulations.

	SMF	DCF
fibre attenuation [dB/km]	0.19	0.5
dispersion coefficient @ 1.55μm [ps/(km*nm)]	16.6	-103.4
dispersion slope @ 1.55μm [ps/(km*nm^2)]	0.06	-0.207
non-linear refractive index [m^2/W]	$2.6*10^{-20}$	$2.6*10^{-20}$
effective core area [μm^2]	69.5	19.0
maximum Raman gain [m^2/W]	$3.2*10^{-14}$	$3.2*10^{-14}$
EDFA noise figure:	≈ 4.5 dB	
attenuation of DCF fibre splices:	0.7 dB	
transmitter:	chirpless, external modulation	
optical receiver bandwidth:	120.0 GHz	
electrical receiver bandwidth:	7.5 GHz	
data signal:	10 Gb/s, NRZ	
back-to-back receiver sensitivity (BER = 10^{-9}):	-37 dBm	

Figure 3 and *Figure 4* show selected results of the numerical investigation. The power penalty shown in these figures is defined as the required increase of receiver input power to maintain a BER of 10^{-9} with respect to the back-to-back receiver sensitivity. The power levels into SMF and DCF, and the Compensation Ratio were varied. CR values between 98% (or 96% in case of the hybrid scheme) and 101% were evaluated. In *Figure 3*, the best case (CR = 99%, Pre-distortion = 40 km) and worst case (CR = 101%, Pre-distortion = 0 km) for the normalised section with pre-compensation are shown. In the worst case, a transmission distance of 1658 km (20 cascades) results in a penalty of at least 4 dB, whereas in best case, 40 normalised sections (3316 km) can be cascaded with a penalty less than 2 dB. For the hybrid-split compensation scheme (*Figure 4*), a CR value of 99% was identified as optimum. After 20 cascades a penalty less than 1 dB is obtained over a wide range of fibre input powers. After 40 cascades, a penalty of less than 3dB is expected in the optimum parameter range.

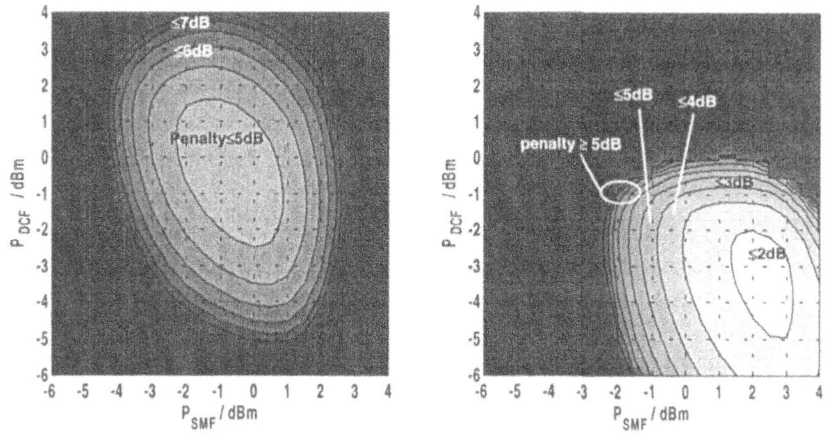

Figure 3. Simulated receiver sensitivity penalties for pre-compensation scheme: 20 loops, CR = 101%, Pre-Distortion 0.0 km (left, worst case); 40 loops, CR = 99%, Pre-Distortion = 40.0 km (right, best case).

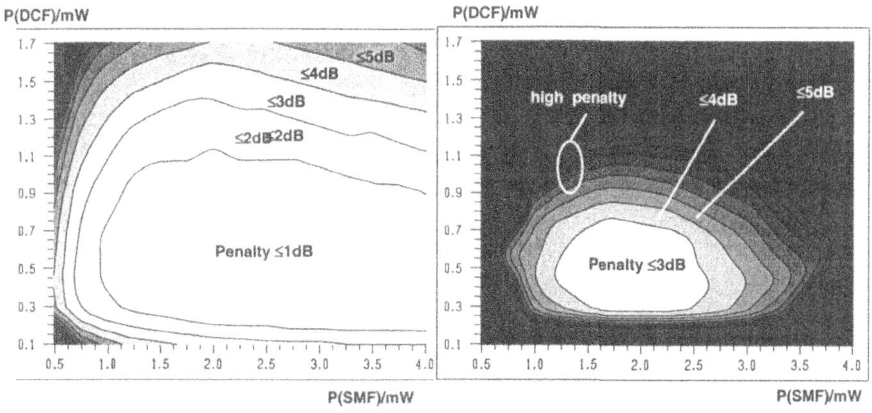

Figure 4. Numerical results for *hybrid-split compensation scheme* for CR = 99% (*best case*) after 20 cascades (left) and 40 cascades (right). Penalty values of more than 5dB are summarized by 'high penalty'.

2.3 Experimental validation of numerical simulations

To verify the promising results of system simulations, fibre loop experiments (s. *Figure 5, Table 1*) were realised. To adjust the compensation ratio, small pieces of additional SMF were inserted in the loop, as well as the operating wavelength was slightly changed. An additional EDFA was used to compensate for the loop attenuation mainly given by the 3-dB coupler and the insertion loss of the loop switch. For the pre-compensation scheme, Figure 6 shows the penalty vs. the number of cascaded normalised sections for different compensation ratios and in the case of 40 km SMF pre-distortion. The penalties were measured for the respective optimum of the SMF/DCF input powers.

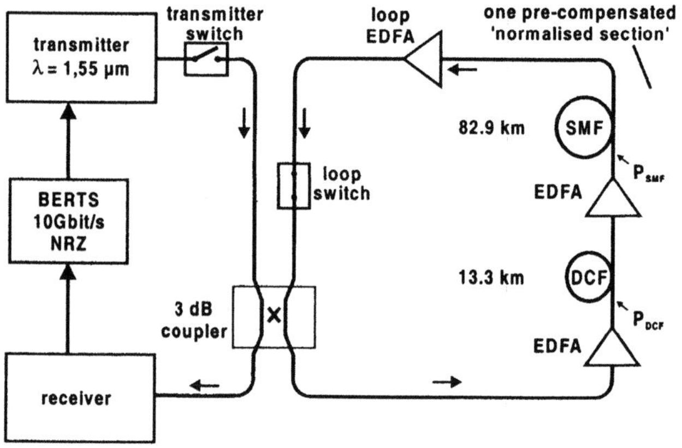

Figure 5. Fibre loop set-up for pre-compensated normalised section.

Without pre-distortion, Figure 6 shows an optimum system performance for a compensation ratio of 99%-100%. In this range, 25 normalised sections (2072,5 km) can be cascaded with less than 4 dB penalty. It can be seen that due to additional pre-distortion a 10 Gb/s NRZ signal can be transmitted over more than 4000 km.

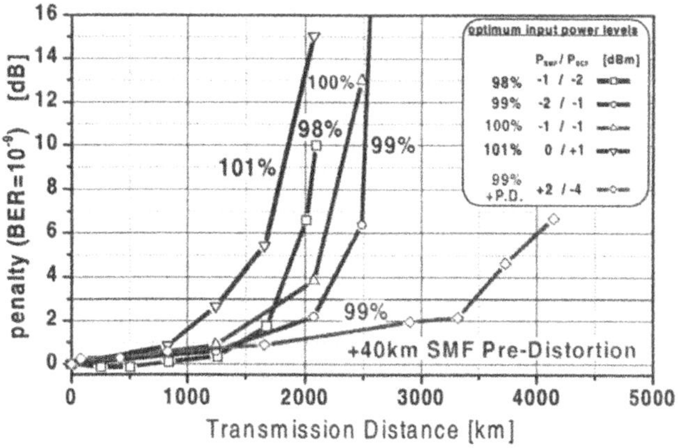

Figure 6. Ultimate transmission distances for the pre-compensation scheme for varying compensation ratio (measurements for 10 Gb/s, NRZ).

Measurement (*Figure 7*) and simulations (*Figure 3.*) show very good concurrence. As it was predicted by simulations, in case of pre-distortion the optimum power region moves toward lower DCF and higher SMF input power. Because the values for the optimum input power strongly depend on the noise dependence of the individual amplifiers, the simulations shown in *Figure 3* also take into account the measured gain saturation of the amplifiers, the measured input power dependence of the EDFA noise figures, and the additional losses due to the connectors in the transmission path.

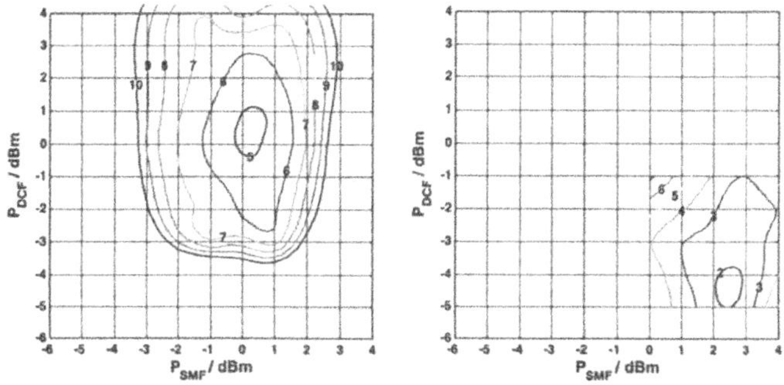

Figure 7. Measured receiver sensitivity penalties for pre-compensation scheme: 20 loops, CR = 101%, Pre Distortion 0.0 km (left, worst case); 40 loops, CR = 99%, Pre-Distortion = 40.0 km (right, best case).

Figure 8 depicts the experimental setup for the hybrid compensation scheme. In this case, 2 transmission sections are realised in the loop.

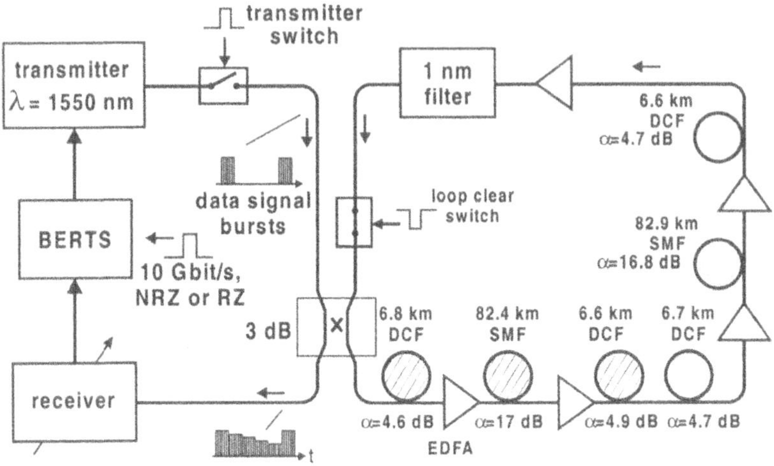

Figure 8. Experimental setup for the hybrid compensation method.

In *Figure 9* the ultimate transmission distances using this hybrid-split compensation technique are shown for different degrees of compensation. The power levels are selected for maximum cascadability, respectively. In the optimum operating range of 99% compensation ratio, 50 cascades (4145 km) result in a penalty of less than 3 dB. Simulations and measurements show very good concurrence.

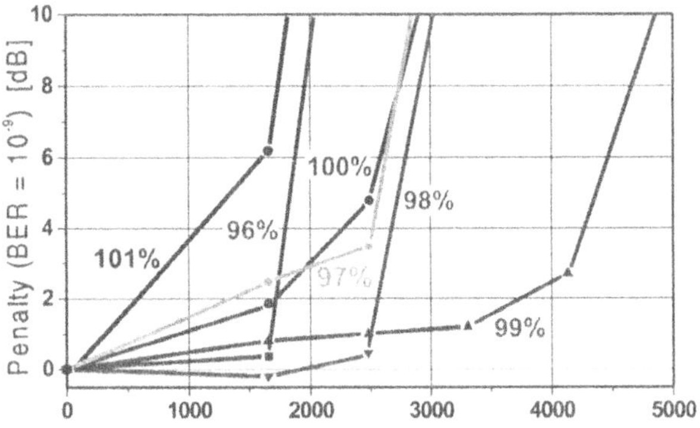

Figure 9. Ultimate transmission distances for the hybrid-split scheme for varying compensation ratio (measurements for 10 Gb/s, NRZ).

3. CONCLUSION

A high amount of system simulations have been performed to investigate the optimum design and operation of transparent, dispersion managed transmission links. Time consuming system experiments have been carried out to verify the results. Evaluation of the numerical and experimental data lead to the following conclusions:

- At moderate transmission distances (up to 1600 km), high variations of input power levels into each fibre segment, and large deviations of the compensation ratio from the ideal value of 99% are tolerated.
- It is important to notice that the ultimate transmission limit of 50 cascades can only be obtained if Polarisation Mode Dispersion is very small along the whole transmission link. (The fibre material installed in the ring showed PMD values of ≤ 0.1 ps /√km) The fibres for such a cascade should therefore be selected very carefully, otherwise the link distance would be limited by PMD.
- With optimised link design, a single channel, 10 Gbit/s NRZ data stream can be transmitted over 4000 km of Standard Single Mode fibre with amplifier distances of 80 km. This assumes sufficiently low values for fibre attenuation and polarisation effects like Polarisation Mode Dispersion, and Polarisation dependent Gain or Loss
- The two Normalised Section designs investigated here have comparable tolerance ranges for the input powers into SMF and DCF, and likewise overlapping tolerance ranges for the degree of dispersion compensation. This means that these designs can be operated with identical parameters, a fact that greatly simplifies the operation of transparent links.
- The optimisation process for the 'pre-distortion' of the pure pre-compensated sections converges towards a 'symmetrical' dispersion map around a dispersion value of zero. In the case of pre-compensation, an addition of a 40 km fibre piece at the transmitter (plus the appropriate DCF piece at the receiver) leads to optimum system performance. The resulting dispersion map is very similar to the 'hybrid' section design, where the symmetrical dispersion map is perfectly realised

Acknowledgement: This work was supported by the European Community under ACTS-DEMON.

4. REFERENCES

[1] E.-U. Scheuing, F.-C. Tischer: *Photonic network design based on reference circuits.* Conference on Optical Network Design and Modelling. Vienna, February 24-25, 1997.

[2] N. Hanik: *Nonlinear Jitter of Optical WDM-Systems due to Cross-Phase-Modulation.* NOC'97, Antwerpen, Proc. 'Core and ATM Networks', pp. 178-181.

[3] G. P. Agrawal: *Nonlinear Fiber Optics.* Academic Press Inc, (1994).

[4] C. J. Anderson and J. A. Lyle, Technique for evaluating system performance using Q in numerical simulations exhibiting intersymbol interference, Electronics Letters, 1994, vol. 30, no. 1, pp. 71-72.

A Simulation Model for an Optical IP Router

F. Callegati*, G. Corazza*, N. Portinaro**, C. Raffaelli*
*D.E.I.S. University of Bologna, Viale Risorgimento 2, 40136 Bologna, ITALY
{fcallegati|gcorazza|craffaelli}@deis.unibo.it - www.deis.unibo.it
**ARTIS Software Corporation, 12 South First Street, Suite 505 – San Jose, CA – 95113-2403
nportinaro@artis-software.com - www.artis-software.com

Key words: optical packet switching, simulation, Internet, Petri Nets.

Abstract: This paper deals with the problem of simulating an optical packet switch for IP traffic (asynchronous variable length packets) relying on wavelength encoding for routing and on delay lines for buffering. A simulation model of this switching architecture based on the use of the Petri Nets formalism is presented, and its implementation by means of the Artifex simulation tool described.

1. INTRODUCTION

The explosive growth of the Internet of the last few years demands for ever higher bandwidth capacity, in particular in the core part of the network. The recent introduction and rapid growth of the Dense Wavelength Division Multiplexing (DWDM) technology provides a platform to exploit the huge capacity of optical fibre, and DWDM optical networks are poised to dominate the infrastructure, supporting the next-generation high-speed Internet backbones. Nevertheless, current applications of WDM focus on the static usage of individual WDM channels, which may not be efficient in supporting IP traffic. The challenge is to combine the advantages of WDM with emerging all-optical packet switching capabilities to yield optical routers capable of guaranteeing full optical data path and throughput in the hundreds of Terabit/s range [1].

Optical packet switching has been addressed by several projects in the last few years [2-4], showing the feasibility of this concept. In order to provide an efficient and flexible platform for IP traffic, we address the scenario of a transfer mode based on variable length packets and asynchronous node operation, according to the "Optical Burst Switching" paradigm proposed in [5]. The adoption of a transfer mode based on long, asynchronous, variable length packets will enable the development of a new generation of optical packet switches which will overcome the limitations of a previous generation of switches specifically designed for fixed length, ATM-like packets.

At present there is very little in the published literature addressing this new transfer mode, and related architectures or performance. New analytical models as well as simulation tools are needed. Simulation tools are particularly important in this class of problems, since in many cases the complexity of the systems in question renders a purely analytical approach very difficult, if not impossible.

This paper begins with the optical packet switch architecture originally proposed in [6] and describes a simulation model of such an architecture. The simulation model has been built within the Artifex simulation environment. Artifex is a commercial simulation tool that uses an extended version of the Petri Nets formalism (Extended Petri Nets) to describe system behaviour. This formalism is well suited to represent telecommunication network problems [7] and the extensions available in Artifex make it a fully powered simulation language, able to process (i.e. simulate) data structures and to incorporate pre-existing or ad-hoc bits of code to perform specific functions.

In this paper we present the switch simulator design and describe in detail the implementation of some of the more significant functional sub-blocks. The aim is to demonstrate the effectiveness of the Extended Petri Nets formalism for this class of problems and to provide the reader with an idea of the simulation capabilities of the model.

The paper is structured as follows: in Section 2, the switching matrix architecture is introduced. In Section 3, the general simulation tool used is outlined, with particular reference to the formalism used to describe events. In Section 4, the simulation model is described in detail, both in general and in more detail choosing some functional sub-blocks as examples. Some numerical results regarding the time required for simulation are presented. Finally, in Section 5, conclusions are drawn.

2. SWITCHING MATRIX ARCHITECTURE

The packet switch here considered is an output queuing switching matrix that relies on the use of fibre delay lines for buffering and output congestion resolution and on wavelength encoding for routing and to achieve optimal exploitation of the buffer. For the sake of simplicity, in the description that follows, each input link is assumed to carry only one wavelength, but the extension to a WDM scenario is straightforward. A block diagram of the switching matrix is shown in Figure 1. This diagram is strictly functional in the sense that there is no intention of addressing implementation issues. The functional sub-blocks are shown as boxes regardless of the way they can actually be implemented. Nevertheless, to the best of our knowledge, any functional block appearing in the architecture is presently feasible with state of the art optical technology.

The switch consists of three functional blocks:

- the packet encoder, which consists of a set of L wavelength converters to access the delay stage in a contention free manner;
- the variable delay stage, which provides multiple stages of delay lines;
- the wavelength selector, to forward a specific wavelength to the addressed output.

On top of these blocks is the control function (not shown in the scheme), which performs the routing function and controls wavelength assignment and packet forwarding.

The packet encoder

When a packet arrives on a switch input link, its header is read by means of an optical detector and processed to determine the switch output to which the packet must be forwarded. On the optical path, each packet is assigned to a wavelength chosen such that no other packet is using the wavelength or is going to use it while the new packet is in the delay stage. Different algorithms can be adopted to achieve optimal use of the wavelength pool; in any case they must be kept simple enough to cope with the very high speed of the switch. In this work it is assumed that the first wavelength is chosen that satisfies the above constraints starting from a fixed wavelength (e.g. the first one). The encoding wavelength is maintained for a packet during the whole switch path.

The variable delay stage

After encoding, the packet enters the delay stage and flows into the fibres of the proper length to be delayed of the required amount. A multistage configuration of the delay lines is proposed to achieve a fined-grained implementation of delay. A base m representation of packet delay is given through the k stages, with stage weights in decreasing order. For a given k, the i-th stage is composed by m fibres that introduce the following multiples of the delay unit D: 0, m^{k-i}, 2 m^{k-i}, 3 m^{k-i}, ... , $(m-1)m^{k-i}$. By properly choosing the delay line at each stage all possible delays between 0 and $D_M=D\,m^{k-1}$ are possible. To achieve the correct composition of the delay contribution, each delay stage is followed by a combining/splitting function, a pool of filters that separate wavelength encoded packets and a space switch to send a packet to the correct FDL in the following stage. The connections between the combiners and the splitters are the critical points of the architecture where it must be assured that no packet jamming takes place on the same wavelength. At these connections only one packet at a time can be present on a given wavelength and this is taken into account by the wavelength encoding algorithm. The values of m and k must be chosen in relation to the required performance. The values are limited by practical constraints and their influence on performance is investigated.

The wavelength selector

This is the output stage of the switching matrix and, at any time, it selects a packet for the correct output. It is built by means of a cascade of optical filters and an optical space switch. The control logic is aware of the time a given packet will leave the buffer and its encoded wavelengths. Accordingly, the control logic sets the space switch to connect an optical filter to the correct switch output.

The control block

The control block provides the routing function and the management of the switch resources. The routing function is based on the detection of the packet header using standard Internet routing procedure and should be suitably performed in advance of the packet arrival. The result of the routing function is the selection of the specific output link to which the packet must be forwarded. This also implies knowledge of the delay that the packet will encounter in the switch, based on the status of the output queue. The next step is to find the wavelength to be used for packet encoding such that the required delay, that is the FDL buffer, is available at that wavelength for the whole packet duration T_P.

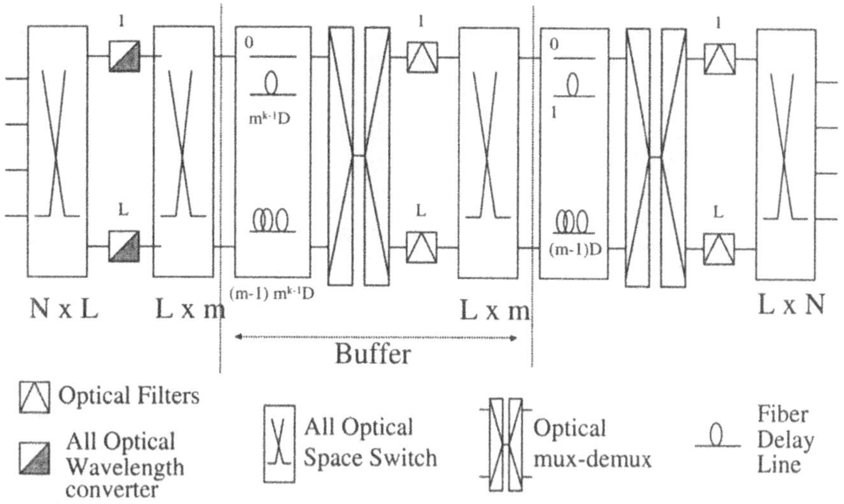

Figure 1. Functional architecture of the switching matrix

3. THE SIMULATION TOOL

Artifex has been used for modelling and simulating the optical switch. Artifex is a commercial tool designed to perform discrete events simulation of systems, and is particularly oriented to the modelling of the logical and dynamic behaviour of telecommunication networks [8]. With Artifex it is possible to model a single network component (e.g. a switch) as well as a complete network. The model can be executed on a graphical simulator to obtain insight into the system behaviour under development, to validate the model itself (i.e. verify if it behaves according to the specifications) and to obtain performance measurements for system optimisation.

Artifex uses a modelling formalism based on Petri Nets. However, in Artifex several features have been added to the conventional Petri Nets formalism, resulting in what is called Extended Petri Nets. It is well known that Petri Nets are an efficient, visual approach to modelling the dynamic behaviour of queues and protocols. In Artifex, a model is defined visually with Petri Nets (by using the Artifex.Editor tool) and translated into a language suitable for simulation.

This approach allows validation at each design step, ensuring that further modelling is always made on safe grounds. Even radical changes can be made in the model in a very short time and the effects of such changes can immediately be simulated with the Artifex.Validate tool.

This Extended Petri Nets formalism provided by Artifex is object-based and allows hierarchical design. It naturally addresses the issue of scaling the complexity of switches and routers as well as end-to-end network protocols.

Artifex models can embed C or C++ statements directly and it is also possible to link external C/C++ code and libraries. This latter feature makes possible the use of C and C++ programming in addition to the Artifex graphical language, for instance to implement algorithms and to perform complex computations on data.

3.1 Basic concepts of Petri Nets

The basic Petri Nets formalism consists of a very small set of symbols. This makes the approach easy to learn. Concepts like parallelism and synchronisation are expressed with a few icons. Petri Nets are defined in terms of three symbols.

- *Place* (represented by a circle) to model states, conditions, events and queues.
- *Transition* (represented by a rectangle) to model state transitions and activities.
- *Arc* (represented by oriented lines) to define cause-effect relationships by connecting places and transitions. Arcs going from places to a transition define transition's input places; whereas arcs going from a transition to places define the transition's output places. A place is often an input to a transition and an output from other transitions -- at the same time.

Current states (i.e. true conditions or occurring events) are represented by drawing a dot within the suitable place. Such a dot is called a *token.*

Petri Nets evolve by moving tokens from an initial state defined by a set of initial tokens until no more tokens can be moved according to the following simple rule: *"When all the input places of a transition contain at least one token the transition can fire. The effect of a transition firing is to remove tokens from some places and to create tokens in some other places. When a transition fires, it fetches one token from each of its input places and releases a new token in each of its output places."*

Some of the extensions to the Petri Nets formalism provided by Artifex are listed in the following [8]:

- tokens can carry data because they can be represented, actually, by data structures;
- it is possible to associate with transitions a specific action (written in C/C++ code) that is executed when the transition fires: the action has access to the content of the input tokens consumed during firing and can modify the content of the tokens that are released in the output places;

- the release of the tokens on the output places can be delayed by a specified amount of time that can be modified dynamically during the simulation, a feature that has been used for modelling FDLs and packet generators.

4. THE SIMULATION MODEL

In this section we describe the implementation of the switching matrix simulation model with the Artifex simulation tool. The model we present here is for a 4x4 switching matrix, with a 3-stage buffer, internally using 10 wavelengths. The reason of this choice is mainly to have a readable overall scheme of the presented model. There are no limitations on the switch size and on the number of wavelengths used internally.

The main goal of this chapter is to show both how the model has been implemented and, by means of some examples, how the Extended Petri Nets formalism has been used for this particular task.

The model has been split into objects. According to the Artifex architecture *object types* are defined and modules performing the same function belong to the same type and are called *object instances*. Objects types and instances have been used here in such a way that the final model resembles the functional scheme of the switching matrix. This is not necessary in principle and also does not optimise the complexity of the model, at least from a graphical point of view. On the other hand, this approach has been chosen because, thanks to the similarity with the physical systems, it makes easier the debugging and development of the model itself.

Figure 2 shows the general model as it appears from the Artifex console. In this simulation model, each packet is a token that will either reside in a place or in a transition while firing. Each packet (token) is actually a data structure carrying information that can be used for processing into the various blocks; for instance packet length, input/output addressing pairs, delay experienced in the FDL buffer etc. Tokens released by transitions carry information updated by the processing, if any, implemented in the transition.

On the left hand side of figure 2, the traffic generators are placed (`IN_x:MM1`). We implemented one generator per input. One single generator could have been used in principle, but in that case the graphic of the model would not have shown the various inputs as different logical entities. It is worthwhile to underline that the overall computational complexity is not affected by having 4 traffic generators instead of one, since the overall number of operations is the same. The traffic generators are followed by a functional block, called `SPLITTER`. Splitters are necessary because the token forwarding between objects is “token broadcast”, i. e. a token outgoing from an object is replicated to all connected objects. The objects of class SPLITTER have the task of splitting packets according to the outputs they are directed to. Actually, a splitter is just a transition with some added C code. The C code, by means of a library function devoted to this task (xx_setaddress), changes the default Artifex behaviour and forces a selective token routing.

There are several splitter blocks in the model. The splitter called SPLITTER_1 forwards the packets towards the delay assignment function (ASS_DELAY). Since packets are assigned a delay accordingly to the occupancy of the output port they are addressed to, the splitter forwards the tokens according to the final destination of the packets they represent.

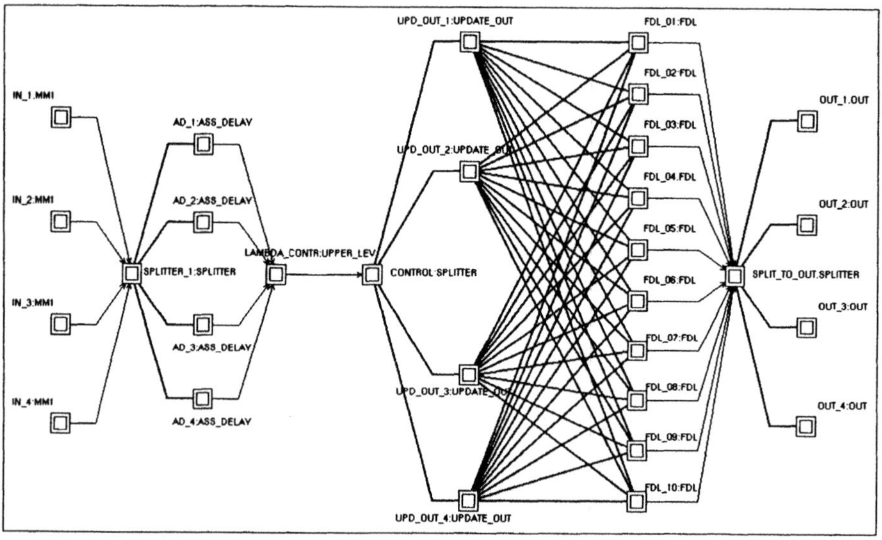

Figure 2. Functional architecture of the switching matrix

The wavelength assignment stage (LAMBDA_CONTR) follows. This is collapsed into one single functional block, as this function is centralised for the whole switching architecture, and is not performed per input or output link.

A further splitter and the stage implementing the FDL buffer (FDL_xx) follows. The buffer is made up of as many blocks (of class FDL) as the number of internal wavelengths. It is worth noting that the splitter CONTROL performs the same function as SPLITTER_1 but according to a different logic. It forwards the tokens to the proper FDL block on the basis of the wavelength they have been assigned to and not on the basis of their output destination. The delay stage has been implemented as a series of blocks and not as one single functional block even though this would have been possible. The approach followed made the debugging and validation of the model simpler. This will appear clearly in the following paragraphs, where the implementation of the FDL stage will be described in detail.

Finally, the output stage follows. In between the wavelength assignment and the FDL buffer, the `UPDATE_OUT` sub-block is present. This block Properly operating the control logic is necessary, since it updates the lists of the output occupancy that are used to assign delays.

To better explain how we used the functions and the modelling scheme as available in Artifex, we will now briefly describe how some of these blocks have been implemented.

4.1 Traffic generator

Let us examine the traffic generator (figure 3). The traffic is random (Poisson process) with exponential packet lengths. For the purpose of this discussion we will call:

- t_i arrival time of packet i: the instant of arrival of the first bit of packet i;
- $T_{i,j}$ inter-arrival time between packet i and j,
- L_i the length in bits of packet i;
- C link speed.

The problem is to generate both inter-arrival times and packet lengths independently, but in a consistent way. If these random variables are simply generated independently, it may happen that a long packet is followed by a very short inter-arrival time. This is not realistic : the bits of a packet are transmitted serially on transmission lines, therefore $T_{i,i+1} \geq t_i + L_i/C$. If this was not true , we would have a situation in which a packet is overcoming the previous one. Therefore, to generate the traffic in a realistic way, we used the technique of generating the traffic in a fully independent way, feeding it in a queue where the packets are served. Since the queue is M/M/1 (Poisson arrivals and exponential service time), accordingly to Burke's theorem, the point process of the departures is still a Poisson process, but now the packets are spaced accordingly to their effective length (since the server must serve one packet at a time). This was implemented as in figure 3. At the beginning a token is placed in place `START`, thus enabling transition `ARRIVAL` that fires with no delay, placing a token in place `QUEUE` and in place `TIME_BETWEEN_NEXT`. Transition `ARRIVAL` also sets the delay in transition `NEXT_PACK`, that fires at exponential rate, re-enabling place start. The queuing packets are placed in service by instantaneous transition `IN_SERVICE`, which is enabled when there is at least one queuing packet and the server is free (place `FREE_QUE` has a token inside). The transition `DEPARTURE`, firing at exponential rate, with different mean with respect to transition `ARRIVAL`, gives to each packet its length and also frees the server (enabling transition FREE_QUEUE). Due to this model, packets are generated as a Poisson process, their lengths are exponentially distributed, and their effective spacing in time is coherent with the lengths.

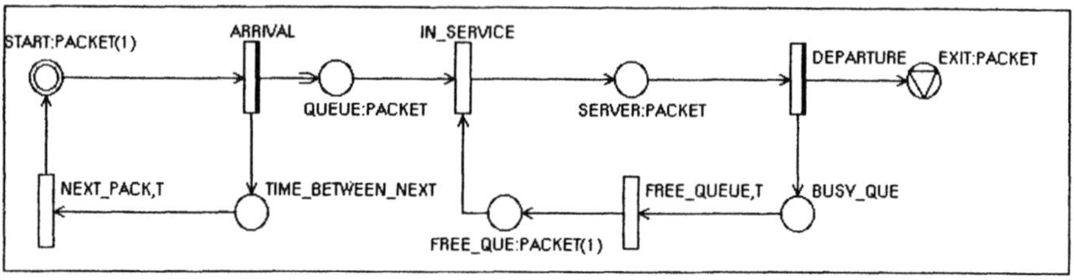

Figure 3. Model of the M/M/1 based traffic generator

4.2 Fibre Delay Line Buffer

The second example presented is the FDL buffer in figure 4. This is definitely easy to implement. The delay a packet must experience is encoded into a token field by the ASS_DELAY functional block, according to the output link occupancy. The delay stage is then just the sequence of as many transitions (FIBRE_xx) as the stages of FDLs. Each transition will fire with a required delay. All the delays will sum up to the overall value required to solve congestion on the outputs. Since the transitions are placed in chain, the firing of one transition on the left automatically enables the following transition rightwards. Each transition also enables the immediate transition UPD_LIST that is used to update the list of time occupancy of the FDL, necessary to the wavelength assignment function. Note that places SIMPLE1, SIMPLE2, SIMPLE3 as well as SIMP_TEMP1, SIMP_TEMP2, SIMP_TEMP3, are connected in cascade mode, they fetch and release the tokens at the same simulation time, since the transitions in between are instantaneous. Therefore they do not contribute to the system evolution, but are necessary to perform the logical function in UPD_LIST.

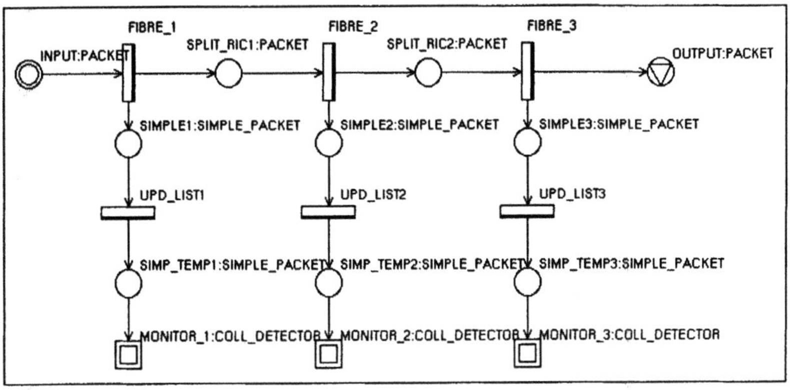

Figure 4. The Fibre Delay Line buffer implemented by means of Extended Petri Nets, including monitoring for wavelength usage (COLL_DETECTOR).

The COLL_DETECTOR blocks are used to verify the correct behaviour of the simulation model and each block is structured as in Figure 5. Simultaneously, the tokens reach place IN_PACKET and the first bit of a packet leaves the fibre delay line. The function contains two transitions: RISE_NEW_COLLISION and SET_TIMEOUT. SET_TIMEOUT transition will fire before (as indicated by digit 1 that follows the name of the transition) and will release a token in place BUSY. It will also set the delay of TIME_OUT accordingly to the length of incoming packet (token).

The transition RISE_NEW_COLLISION is enabled when both places IN_PACKET and BUSY contain a token. The occupancy of place BUSY represents the period of time required by a token to cross one of the combiner/splitter pairs that connect the delay stages (see Figure 1). In these critical points only one packet at a time must be present and encoded at a given wavelength. If a token gets place IN_PACKET while the previous one is still in place BUSY, then RISE_NEW COLLISION fires and releases one token in place COLLISIONS. The number of tokens in this place represents how many times there have been collisions and the algorithm failed.

This is probably the most evident example of the use of the validation capabilities of the Extended Petri Nets formalism in this model. At the end of a simulation, or even while it is still running, the correctness of the wavelength assignment algorithm can be visually checked by simply checking whether there are tokens in place COLLISIONS or not.

This monitoring function is also the reason why we implemented as many delay blocks as the number of available wavelengths. By doing this, you can check that no collisions occur on each internal wavelength. This would not be possible (or would be anyway more complex from the implementation point of view) in the case of a single delay block crossed by packets encoded on any wavelength.

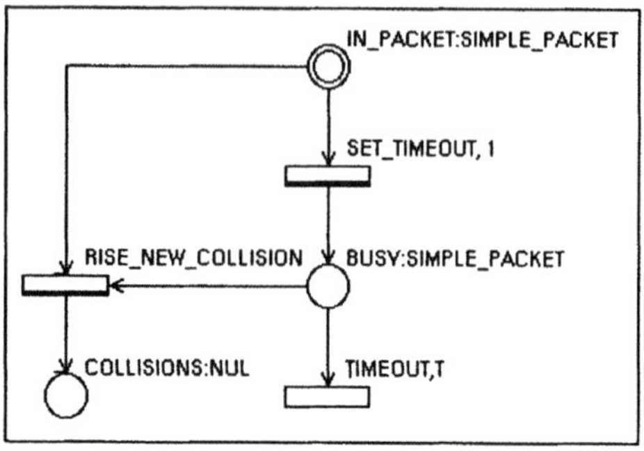

Figure 5. The implementation of the functional block COLL_DETECTOR in the FDL buffer

4.3 Simulation Performance

An in depth analysis of the performance of this switching architecture is beyond the scopes of this work and the interested reader can refer to [6] for a detailed description of such results. On the other hand here we focus on giving some results on the simulation capabilities of the model described.

We ran a series of simulations within two different environments. We first simulated a 4×4 switch with 10 wavelengths and then a 16×16 switch with 26 internal wavelengths. The FDL buffer was made of 3 stages with 16 fibres per stage. The average packet length was chosen to be L=500 bytes and the unit of delay of the buffer D=30 bytes. With these values the maximum delay achievable was equal to 4095 D that is 4.1 μs assuming a link speed C=2.4 Gbit/s. For each case we simulated about 10000, 100000 and 1000000 packets using a Pentium II personal computer, with 256 Mbytes of RAM and equipped with the Linux operating system.

The following tables show the run time needed for each simulation. They also show the packet loss probability measured and the number of lost packets thus giving an idea of the degree of confidence which can be achieved. The simulation time was obtained using the Linux `top` command, measuring the effective CPU time used by each simulation.

# of simulated packets	4×4 Simulation Run Time	4×4 # of Lost Packets	4×4 Packet Loss
10435	13”	59	5.654E-03
106146	2’ 20”	734	6.915E-03
1058356	23’ 26”	7326	6.940E-03

Table 1. Numerical values of the simulation time for the 4×4 switch for various values of the number of simulated packets. The measured packet loss is also shown.

# of simulated packets	16×16 Simulation Run Time	16×16 # of Lost Packets	16×16 Packet Loss
10034	14”	1	1.000E-04
102008	2’ 14”	52	5.100E-04
1011010	140’ 1”	507	5.010E-04

Table 2. Numerical values of the simulation time for the 16×16 switch for various values of the number of simulated packets. The measured packet loss is also shown.

5. CONCLUSIONS

In this paper we presented an overview of the simulation model of an optical packet switch exploiting wavelength multiplexing for routing and delay lines for buffering of packets. The paper gives an overview of Artifex, the simulation tool that has been used, and presents the simulation model with some numerical results regarding the simulation time of typical system configurations. Furthermore, the implementation of the most significant functional sub-blocks has been described in detail. We believe this model can be of help to researchers facing similar tasks and can be used as a reference for the development of more complex models.

ACKNOWLEDGEMENT

The authors wish to thank Mr. Alessandro Botticelli for his help in developing the simulation model and running the simulations.

REFERENCES

[1] B. Bostica, F. Callegati, M. Casoni, C. Raffaelli, “Packet Optical Networks for High Speed TCP-IP Backbones”, *IEEE Communication Magazine*, Vol. 37, No. 1, pp. 124-129, January 1999.

[2] F. Masetti et al., “High speed, high capacity ATM optical switches for future telecommunication transport networks”, *IEEE Jornal on Selected Areas in Communications*, Vol. 14, No.5, pp. 979-999, 1996.

[3] L. Chlamtac et. al., “CORD: contention resolution by delay lines”, *IEEE Journal on Selected Areas in Communications*, Vol. 14, No. 5, pp. 1014-1029, 1996.

[4] P. Gambini et al., ``Transparent optical packet switching: network architecture and demonstrators in the KEOPS project", *IEEE Journal on Selected Areas in Communications*, Invited paper, Vol. 16, No. 7, pp. 1245-1259, 1998.

[5] J. Turner, “Terabit Burst Switching”, *Journal of High Speed Networks*, Vol. 8, No. 1, pp. 3-16, 1999.

[6] F. Callegati, G. Corazza, C. Raffaelli, "An Optical Packet Switch with a Multi-stage Buffer for IP Traffic", *Optical Networking*, A. Bononi ed., pp. 300-311, Springer-Verlag, London, UK, 1999.

[7] .M. Baldassari, R. Benso, R. Pane, "Modeling WDM Protocols and Networks Using the Artifex Environment", Proc. *SPIE – All-Optical Networking 1999: Architecture, Control, and Management Issues*, Boston, MA, September 1999.

[8] *The Artifex Language*, by ARTIS Software Corporation, Torino, August 1997.

Distributed Router Architecture for Packet-Routed Optical Networks

Michael Düser, Eugene Kozlovski, Robert I. Killey and Polina Bayvel
Optical Networks Group, Department of Electronic and Electrical Engineering, University College London, Torrington Place, London WC1E 7JE, UK, Tel.: +44-20-7679-3843, Fax:+44-20-7388-9325, e-mail: mdueser@ee.ucl.ac.uk

Key words: Optical Networks, Packet Networks, IP over WDM, MPLS, Burst Switching

Abstract: A new proposal for an optical packet-routed network based on a distributed router architecture in a WDM network is described. Buffering, scheduling and wavelength assignment functions are carried out in routers located at the edge of the network. The destination address is encoded by a given wavelength and allows wavelength-routing in optical core routers. This approach avoids processing and additional buffering delays in core routers. A low packet loss rate and a low latency are critical parameters to guarantee QoS for a particular class of packets and must be carefully considered when IP traffic is directly transmitted over a photonic network. This network architecture requires new hardware elements and these are discussed including a low-crosstalk, large port-count free-space grating router.

1. INTRODUCTION

With the projected annual increase of 60% in data traffic [1], solutions are required to combine the advantages of the network packet layer with the high-bandwidth optical transmission in the physical layer. To date the proposed solutions concentrate on current protocols and transmission standards namely ATM and SDH, such as in proposals of IP-over-ATM [2] and packet-over-SDH transmission [3]. Whilst these simplify the protocol stack, limitations in scaling to a large number of routers remain. The advantage of connection-oriented protocols such as ATM is their provision of quality of service (QoS) awareness, usually not considered in connectionless protocols like IPv4.

These problems can be overcome using a new protocol proposed by the Internet Engineering Task Force (IETF) in 1997 [4] known as multiprotocol label switching (MPLS) in combination with optical routing. MPLS, like ATM lies between the network and the link layer but, in contrast to ATM, can operate with variable packet lengths, and directly with IP, avoiding the necessary adaptation of packets to ATM cells.

In this paper we describe an MPLS-based optical network architecture that does not rely on processing and buffering in the core nodes as recently suggested [5]. It is proposed to distribute the routers to the edge of the network, equipped with electronic processing, scheduling and buffering. With packets aggregated to flows, virtual lightpaths are established via MPLS between the ingress and the egress router of the network. MPLS can be additionally used to interconnect the core networks of network operators.

2. MULTI PROTOCOL LABEL SWITCHING (MPLS)

Although MPLS is not a defined standard, it is widely discussed in the Internet Engineering Task Force (IETF) [4] with the following main functions:

- *Short label*

Since MPLS is designed to allow a hop-by-hop routing, it is sufficient to use a short label instead of a full header. This makes it particularly suitable for WDM because the MPLS label with a length of 32 bits [6] (Fig. 2) gives a reduction of a factor of 10 compared to the 320 bits header envisaged for IPv6. With this type of header it is possible to distinguish more than 1 million destination addresses, 8 classes of service, and to identify the bottom label of a label stack. The time-to-live field (TTL) allows a maximum number of 256 hops before the flow is discarded. As labels can be attached to any packet, MPLS can deal with variable length packets such as IP packets (max. 64 Kbytes payload). This is different from the ATM approach with short, fixed size ATM cells (53 byte).

Address label (20 bit)	Class of Service (3 bit)	Bottom of stack (1 bit)	Time to live (8 bit)

Figure 1. Example for a 32-bit MPLS header [8]

- *Hierarchical label stacking*

MPLS allows to stack several labels on top of each other before sending the data flow from the first edge router. Labels are organised as a stack (last-in-first-out [LIFO]) so that the top label determines the next hop. This allows for temporary labels on a MPLS flow, important in case of re-routing.

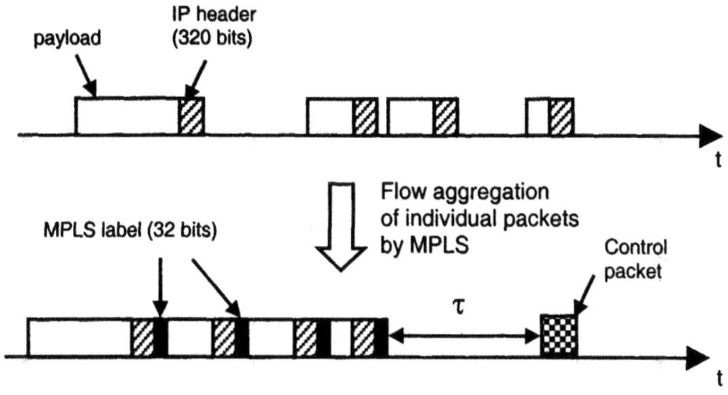

Figure 2. Aggregation of a MPLS flow from single IP packets by adding MPLS labels (32 bits) plus an additional control packet with an offset time τ proceeding the flow

- *Virtual circuit emulation using label-switched paths (LSPs)*

Like ATM, MPLS sets up virtual paths between edge-routers, which are called label-switched paths (LSPs), reserving a particular wavelength for the transmission.

- *QoS routing*

Routing is carried out according to the class of service, i.e. packet flows originating from the same edge router for the same destination but containing different classes of are forwarded on different LSP. Transmission on different wavelengths is therefore inherently supported by MPLS. The implementation can either use a Resource Reservation Protocol (RSVP), or a Label Distribution Protocol (LDP) [7].

- *Failure detection and restoration*

Situated between the network and the link layer, MPLS does not operate at the physical layer of the network and will therefore provide restoration after measures on the physical level are taken. Since MPLS controls the router and hence the release of flows into the network, a fast forwarding of the failure information (tens of ms) to the MPLS control is essential to start re-routing of LSPs [8].

3. NETWORK ARCHITECTURE - A BURST-SWITCHED OPTICAL NETWORK

Currently, wavelength-division multiplexing (WDM) is used for high capacity point-to-point links connecting electronic network routers. Although electronic routers can route and forward packets, their scalability is limited beyond 1 Tbit/s [9].

The ideal infrastructure would evolve from a static wavelength-routed optical network with fixed lightpaths to an all-optical connection-less network resembling an IP network. It is assumed that switching timescales are reduced from months or days for the static model to nanoseconds in the case of a network forwarding at the packet level, in response to load changes.

The principle of burst-switching in the context of WDM networks has been investigated [10] [11], but to date has not been applied to MPLS and optical routing. The principle of this approach is to aggregate the packets to flows (Fig. 2) in the edge routers to reduce the number of wavelength-switching operations. Due to the time required to tune the laser diodes or wavelength converters to the appropriate wavelength the achievable switching speed is limited. Since MPLS operates on a flow-level, and not on a packet-level, packet aggregation is essential. MPLS has the advantage of being quasi-circuit switched. In this architecture, packets are sorted by traffic type and destination in queues in the edge routers. In this scheme we assume 3 classes of service:

- Premium: Traffic with low packet loss rate (PLR) ($< 10^{-9}$ for traffic load 0.8 as in ATM) and nearly no latency (quasi circuit-switched)
- High Quality: Packets with specified PLR and a specified maximum latency
- Best Effort: Regular IP packets with high tolerance towards packet loss and latency

The length Q of the queue is a critical parameter and must be investigated in the context of the proposed MPLS architecture. The size Q of a queue required to store incoming packets is a function of the queueing delay τ_{dq} and the bitrate b of the network:

For example, at a bit-rate of 10 Gbit/s a buffer size of 1.24 Mbyte is required to buffer the traffic passing through in 1 ms. The queues are emptied either when the size of the queue reaches a given threshold or a time-out signal is sent to guarantee a specified acceptable delay. Large memory and processing capacity are concentrated in the edge router, whilst the optical routers in the core act as a large, distributed routing network. The advantages of this scheme over other architectures are as follows:

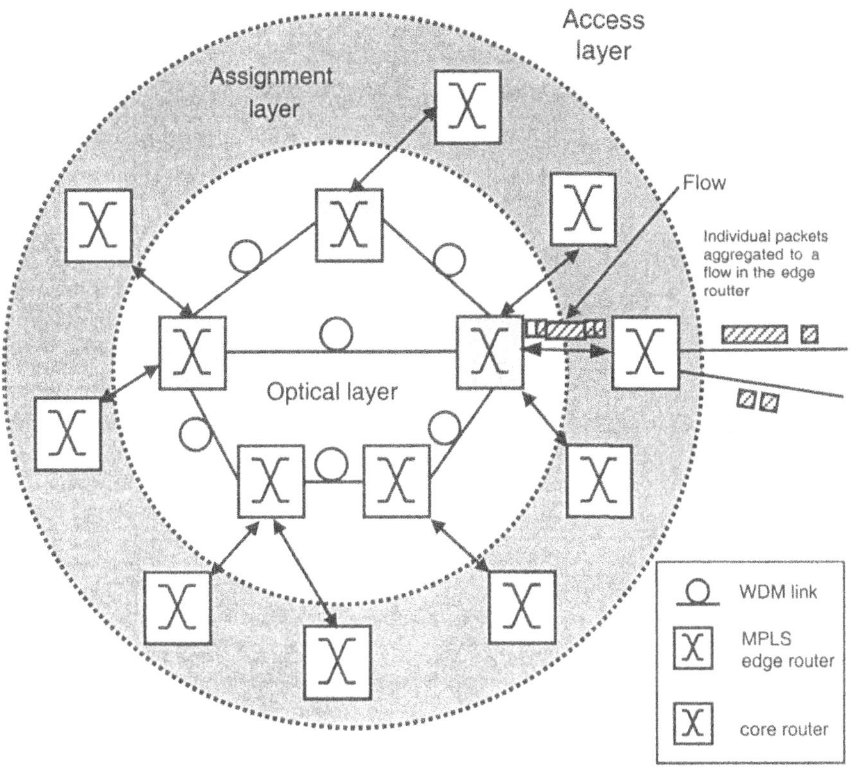

Figure 3. Principle of the proposed network architecture applying burst switching

The millisecond timescale for sending flows through the network allows sufficient time for wavelength selection in transmitter lasers and wavelength converters. Working at the flow level and establishment of short-term connections between edge-routers allows compatibility with the MPLS protocol for the transmission over and control of the optical layer.

The packet latency is determined by the queueing delay prior to release of the packet flow into the core network and directly proportional to its length Q. Once a flow enters the core network it experiences only the propagation delay. Thus queue "threshold" and "time-out" determine the QoS for a particular type of packet.

The edge routers in the distributed router architecture (Fig. 3) connect the access with the optical core network. In addition to the processing, scheduling and buffering these must carry out the functionality of mapping incoming packets to the optical layer of the network. This is done in two steps (Fig. 4):

(1) The packets are aggregated into MPLS flows. This is carried out by reading and processing of the entire packet header information (320 bits for IPv6), the scheduling of the packet and the storage in the correct output queue. This is carried out in the electronic domain where sufficient memory and parallel processing capacity are available to cope with optical line speeds ≥ 10 Gbit/s. The time a flow spends in the buffer depends on the specified maximum delay for a given class of packets and will be on the timescale of < 1 ms for High Quality type traffic.

(2) The second function needs to allocate flows and reserve bandwidth from the input edge router to output edge router through the core network. This is done by matching the address of a flow to the wavelength so that within the core network no more header lookups are required.

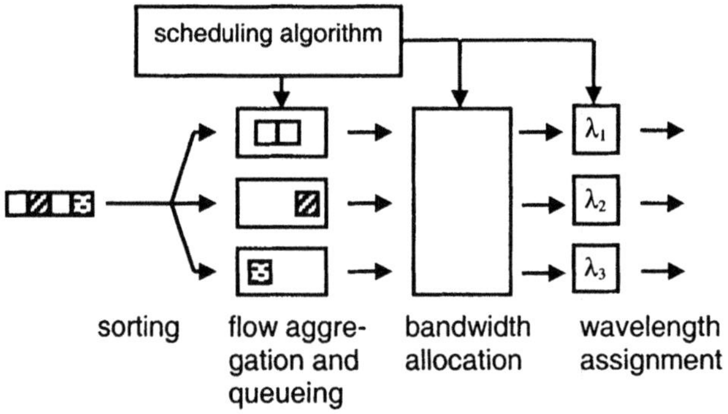

Figure 4. Treatment of packets in the proposed network architecture

The proposed architecture is based on an optical burst-switching (OBS) architecture in combination with MPLS for operation of the network and transmission between edge routers. The scheme operates on three different layers specified in table 1:

- *Access layer*

Packets with different input formats, but mainly originating from IP applications enter the edge routers. MPLS is not restricted to a particular format but is designed to operate with different protocols such as ATM, IP, etc.

- *Assignment layer*

The trade-off is between the length of the queue and the overprovisioning of bandwidth, i.e. the number of wavelength required in the network. The blocking probability or delay can be reduced by providing more wavelengths channels, but in addition to more hardware this increases the network control and management complexity.

Table 1. Reference model for the proposed network architecture using burst-switching

Layer	Functionality
Access layer	Incoming packet: IP, voice, video
Assignment layer	1. Synchronisation 2. Incoming packet header detection and processing 3. Data rate adaptation and convergence 4. MPLS header writing/removal 5. Scheduling 6. Packet aggregation to form flows whilst packets are queued according to packet class and destination 7. Wavelength assignment (layer 3 routing) 8. Bit-rate control
Optical layer	Switching (layer 2) Transport 3-R regeneration

The number of queues N_{queue} that need to be controlled per edge router is determined by the number of edge routers N_{edge} and the number of classes C supported by MPLS:

$$N_{queue} = N_{edge} \times C$$

Assuming a network with 20 core routers and 10 edge routers connected to each core router (N_{edge} = 200) and 8 classes of service as envisaged for MPLS (C = 8), this would result in 1600 queues to be operated by each edge router. In the case of 10 Gbit/s system this requires a capacity of approximately 2 Gbytes per ms of traffic that needs to be stored when the queues are operated independently.

The key function of MPLS is shown in Fig. 5. As MPLS is situated between layers 3 and 2 in the network, it can operate on top of SDH (backward compatibility) or digital wrappers (future component), or it can even run directly on the WDM domain. In this case network failures must be reported in the timescale of tens of milliseconds to the MPLS level to allow fast restoration and re-routing in the network.

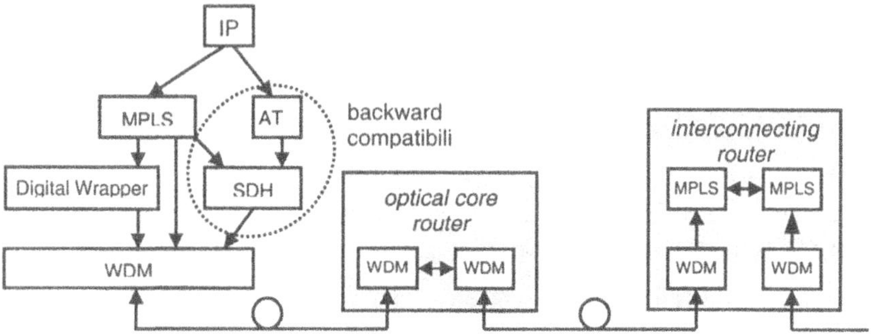

Figure 5. Protocol layers used whilst a packet is sent via different core networks

- *Optical layer*

Transmission and routing of flows is carried out in the optical domain by a core wavelength router where no optical buffering and processing functions are required. However, network operation without processing and buffering in core nodes requires a new scheduling algorithm to provide QoS for the different types of classes for packets in the network. Flows buffered in queues are assigned to a free wavelength, equivalent to the bandwidth reservation.

The most suitable reservation scheme is JET (Just-Enough-Time) [13]. A list of available wavelengths is stored in association with a central scheduling algorithm from which each router obtains free wavelengths as required. The control packet reserves a particular bandwidth for a fixed duration allowing a definite forecast when the wavelength will be available again. JET is important for QoS awareness in the network by assigning different priorities to the flows buffered in the edge router. Priorities are associated with different blocking probabilities and different queuing delays, i.e. the lower the priority, the higher is the associated blocking probability and the queuing delay.

4. CLASS OF SERVICE DIFFERENTIATION BY LIGHTPATH AND DISTRIBUTION TREE ROUTING

Wavelength routing allows bypassing buffering and scheduling at the core routers by providing as many wavelengths as required for the complete contention resolution in each link of the core network.

In the proposed architecture, IP traffic arriving at the ingress router is buffered and scheduled electronically and each aggregated flow is attached an MPLS label and then assigned a wavelength to route the flow along the entire path to the destination. At the expense of wavelength overprovisioning, this eliminates the problem of packet header processing at each intermediate node. Depending on the QoS requirements for a particular traffic class, two methods for wavelength assignment with different amounts of overprovisioning are considered in this section.

1. *Lightpath assignment.* This type of assignment provides direct connections between source and destination and requires no processing or buffering in the optical domain [12]. Therefore, lightpath assignment is preferable for accommodating real-time traffic. Each lightpath is assigned a unique wavelength along all the links between edge routers. The wavelength can be the same in every link of the lightpath or it can vary from link to link along the path in case of wavelength conversion [13]. Let V be the set of nodes, $|V| = N$ and let the set of lightpaths be

$$P=\{p_{ij}(s_i, d_j) \mid i = 1,\dots, N,\ j = i,\dots, N\},$$

where s_i is source node i, d_j is destination node j, p_{ij} is the lightpath between s_i and d_j. Thus, total number of unidirectional lightpaths in the network is $|P| = N(N\text{-}1)$.

2. *Distribution trees assignment.* This type of assignment is provided by the set of distribution trees

$$T = \{t_j(s_i, D_i) \mid i = 1,\dots, N\},$$

where s_i is source node i, $D_i = V \setminus s_i$ is the set of destinations for the source s_i, t_j is the tree routing packets from s_i to the set D_i. The total number of trees is $|P| = N$.

Thus, a single wavelength is assigned to all the edges of each distribution tree t_j which can route traffic from source node s_i to all the destinations of the core network [14] (see Fig. 6 for an example of lightpath and tree assignment in a 4-node network).

Whilst, with the distribution tree approach, the wavelengths are shared among different destinations for each source, packet contention at the intermediate nodes is avoided, as each tree is assigned a unique wavelength along its edges. Packet buffering and scheduling is carried out in the source edge-routers, with packets being transmitted only when the required wavelength becomes available.

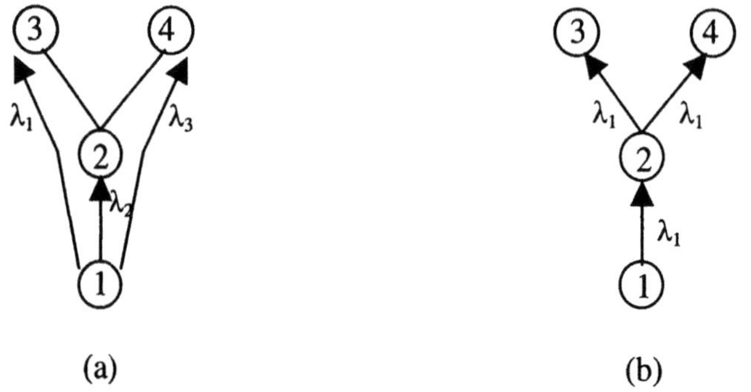

Figure 6. Difference between lightpath and distribution tree assignment in a 4-nodes network. (a) Three wavelengths are required to connect node 1 with the other nodes by lightpaths. (b) One wavelength is required to connect node 1 with the other nodes by a tree as packets routed from one source edge-router to different destination edge-routers share the same wavelength.

The algorithm used to build distribution trees is as follows.

(1) For every source node, a shortest-path tree spanning all the others nodes is calculated.

(2) Since for real network topologies there is usually more than one way to span the nodes with shortest paths, a set of all the possible shortest-path trees is built for a given source node.

(3) All the trees in the set are consecutively considered as the distribution tree for a given source node, and a shortest path tree is chosen as distribution tree if, and only if, the number of wavelengths in the most loaded link of the network is lower than using any other tree in the set.

(4) Steps (2) – (3) are repeated for all the nodes in the network. As a result, distribution tree is assigned to every node and wavelengths are reused so that packets coming from the same source through the given link share the same wavelength.

Consider any node $s_i \in V$. In order to provide unidirectional logical connections to all the other nodes, it must originate N-1 lightpaths, and therefore, requires exactly N - 1 transmitters. At the same time, in order to provide logical connections with all the other nodes by distribution trees, fewer transmitters are required because for each source node $s_i \in V$, several destinations can share the same wavelength on a given fiber, depending on the network topology. Therefore, it can be shown that the total network load, i.e. the number of wavelengths in the most congested link, can be significantly reduced by means of distribution trees compared to lightpaths.

However, the distribution trees impose longer delays on traffic delivery because packets destined to different nodes can only be carried serially along those edges, which share the same wavelength. Additionally, low-level routing is required at each intermediate node in order to route packets to a particular destination along the tree.

Nevertheless, distribution trees represent a trade-off between the number of wavelengths required and the queuing delays at edge routers, and can be used to accommodate lower-priority traffic. Preliminary calculations show that use of distribution trees can reduce the wavelength requirement in such networks as NSFNet, ARPANet and USNet by up to 40% compared to lightpaths. The number of bits required to code packet destination along the tree can be derived from the network diameter. This number can be shown to be almost the same for real network topologies and varies from 2 to 4 bits for the above-mentioned networks.

Hence, considering that packet class-of-service (CoS) field is coded by the assignment type, the entire MPLS header can be mapped into the combination of the wavelength code and a small number of forwarding bits allowing fast switching in the core routers (Fig. 7).

Therefore, distribution trees allow for buffer-free packet delivery with only a small forwarding label to be processed in the optical domain. Another advantage of distribution trees is in their ability to provide direct multicast connections without queuing delays, as in the case of multicast, packets can be transmitted simultaneously [14].

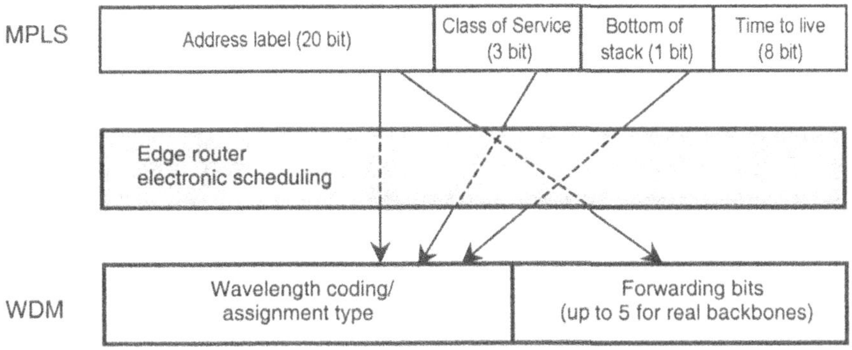

Figure 7. Mapping MPLS header to WDM domain.

5. EXPERIMENTAL NETWORK ARCHITECTURE

The experimental investigation of the proposed network architecture requires a packet generator and analyser, a fast tuneable laser or a set of switchable transmitters and/or wavelength converters, and a large-port optical router. Transmission of packets can also be studied by integrating the label generation and error detection in a single set-up (Fig. 8). Long distance and multi-hop transmission of packets can be studied using a recirculating fibre loop.

The generation and analysis of optical packets at line rates of 10 - 100 Gbit/s throughout is carried in a computer used for the control of the set-up and aggregation of flows. According to the destination address and class of service of the flow a free wavelength is selected and the pattern is simultaneously transferred to the pulse pattern generator (PPG), to modulate the cw wavelength carrier of the multi-wavelength transmitter. The wavelengths carrying flows are routed by the free-space grating router (expected to be at ns rates). The wavelength tuning delay could be reduced by using a multi-wavelength array rather than a tuneable laser.

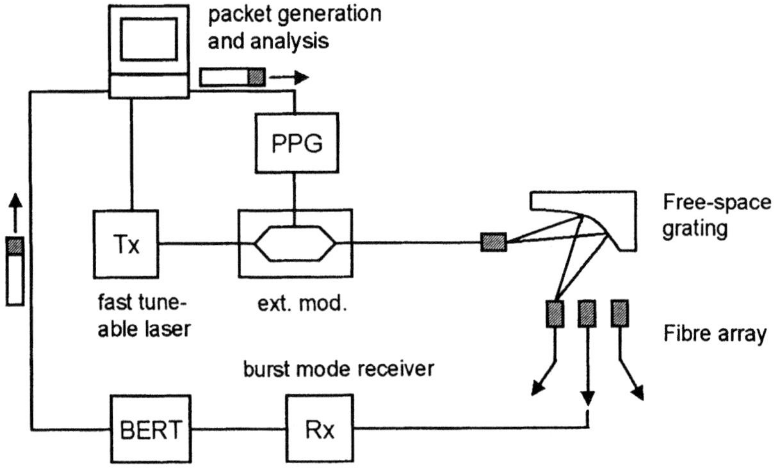

Figure 8. Free-space grating router in DEMUX configuration for back-to-back measurements for distributed switching

Optical switching matrices suffer from problems in the scalability for large numbers of ports [15] due to crosstalk and high insertion loss of the elements. Free-space routers overcome the limitations of waveguide-based routers as they inherently can achieve lower crosstalk values and a free-space grating device (Fig. 9) with < 40 dB crosstalk and better than 3.7 dB fibre-to-fibre loss designed for 100 ports [16] and operating at 10 Gbit/s was recently demonstrated [17]. This achieves the required switching matrix by distributing the switching functionality to the edges.

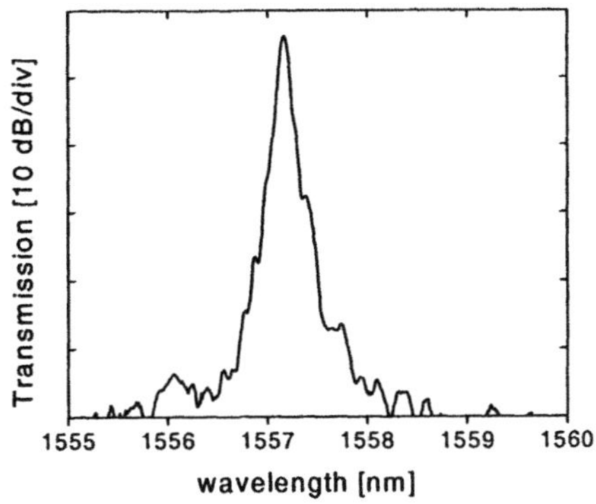

Figure 9. Measured passband of free-space grating demultiplexer [17]

Other key elements are:

- *Burst mode receiver*

Packet error detection requires the implementation of burst mode operation. This requires synchronisation with incoming optical packets at nanosecond timescale. Specifically a separate synchronising pattern is required to align the incoming data stream with the detector reference clock. Use of a burst mode packet receiver will allow the simultaneous measurement of queueing delay, PLR and traffic load. This will allow defining the design rules for a given network.

- *Packet loss, delay and traffic load*

While the BER is the correct measure of the performance of a physical link, QoS parameters such as PLR and packet delay are additionally important for the evaluation of network performance. This requires a three-dimensional surface plot for the graphical representation of queueing delay, PLR and traffic load. The graph can be used in two ways: Either for plotting simulation results showing the performance resulting from a given network topology, or QoS parameters can be specified by a given mask whose values may not be exceeded by the real network.

- *Number of hops*

An important parameter in the design of optical transmission networks is the scalability in terms of number of physical hops. The number of hops a packet can undergo in a photonic network is limited by distortions originating both from the fibre transmission characteristics and distortions added by the optical router, such as crosstalk and insertion loss. The eye closure for a randomly connected network (Fig. 10) decreases linearly with

the number of cascaded channels. Transmission through 12 routers an additional 3 dB penalty is introduced to the system.

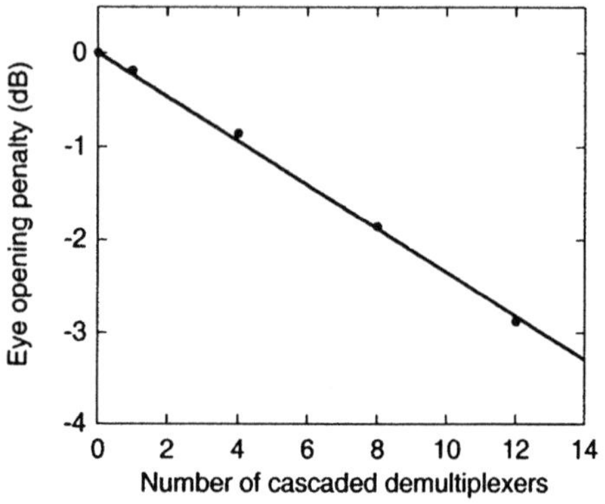

Figure 10. Calculated eye closure penalty with transmission through cascaded demultiplexers, with passband T (dB) = - 0.014 $(f - f_0)^2$

As bit-rates per channel increase, and optical power is increased to maintain adequate signal-to-noise ratios with large inter-amplifier spans, the nonlinear effects of the transmission fibre cause signal distortion, which must be minimised by optimisation of the dispersion map. One technique is undercompensation of the fibre dispersion to avoid pulse broadening due to self-phase modulation (SPM). Fig. 11 shows the amount of under-compensation, achieved with lumped anomalous dispersion at the receiver, required in 10 Gbit/s NRZ format transmission over standard fibre (D = 17 ps/nm·km), with post-compensation in each 80 km span, and 8 dBm launch power into each span. Without the residual dispersion, the eye closure penalty is > 6 dB after 10 spans, while with undercompensation, it is < 1 dB. Although this technique is very effective at extending transmission distances, the optimum amount of residual dispersion is dependent on the transmission distance, and the rapid switching of signals over different routes in burst switched networks will require careful design of the dispersion maps or the development of rapidly variable dispersion compensators.

- *Network topology - total number of edge and core routers*

The number of hops is also related to the number of nodes present in the core network. The proposed burst-switching networks need to be carefully controlled to use bandwidth efficiently and to detect and resolve conflicts. This functionality must be embedded in the scheduling algorithm and will be restricted by the total number of nodes and the connectivity.

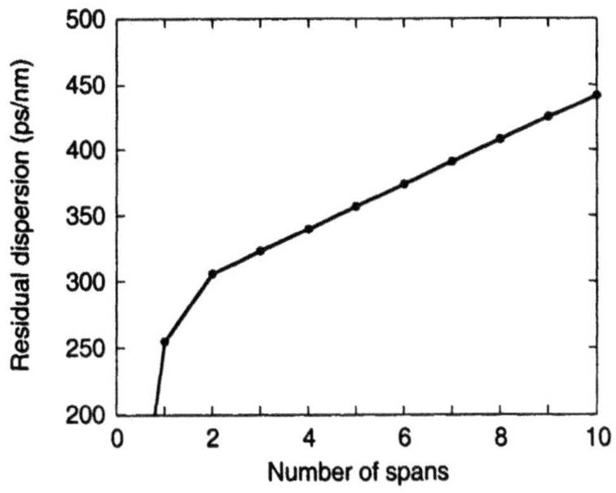

Figure 11. Calculated residual dispersion required to minimise self-phase modulation distortion in multi-span 10 Gbit/s transmission over standard fibre, with 8 dBm launch power

As signalling is required, the round-trip-time (RTT) of control packets limits the size of the network. An additional parameter in the course of the investigation is the maximum number of edge routers that can be connected to the optical core network.

- *Length of burst and queuing delay*

The system performance must be investigated as a function of the length of the bursts and the queuing delay. The trade-off for a reduction of the PLR is an increasing delay because flows have to be stored in memories for longer before being released into the network. Finding the optimum balance between length of burst, queuing delay and PLR is key for the optimisation of optical packet-routed networks.

6. CONCLUSIONS

The advantages of integrating packet transport into the optical domain using the MPLS protocol have been described. Aggregation of packets into flows at edge routers allows the set-up of virtual connections between edge-routers that are passively wavelength-routed inside the network. The bandwidth offered by the large number of channels available in WDM can be efficiently exploited by MPLS, which allows to efficiently set-up different virtual connections for different classes of service whilst utilising the quasi-circuit switched nature of wavelength-routing.

ACKNOWLEDGEMENTS

The authors would like to express their gratitude to Prof. J. Crowcroft (UCL), Dr. R. J. Gibbens (Cambridge), and Prof. S. Ferguson (Marconi Communications) for invaluable comments. Financial support from Marconi Communications, UK EPSRC, HRLD Foundation, UCL Graduate School and the Royal Society is gratefully acknowledged.

REFERENCES

[1] P. Lagasse et al., "Roadmap towards the Optical Communication Age - A European view by the HORIZON project and the ACTS Photonic Domain", Ghent, June 1998

[2] P. Dumortier, "Toward a New IP over ATM Routing Paradigm", IEEE Comm. Mag., Jan. 1998, pp. 82 – 86

[3] J. Carlson, E. J. Hernandez-Valencia, N. Jones, P. Langner, "PPP over Simple Data Link (SDL) using raw lightwave channels with ATM-like framing", IETF draft draft-ietf-pppext-sdl-pol-00.txt, June 1999

[4] R. Callon, G. Swallow, N. Feldman, A. Viswanathan, P. Doolan, A. Fredette "A Framework for MPLS", IETF draft , draft-ietf-mpls-framework-05.txt, Sep. 1999

[5] F. Callegati, K. C. Cankaya, "Design Issues of Optical IP Routers for Internet Backbone Applications", IEEE Comm. Mag., Dec. 1999, pp. 124 – 128

[6] T. Li, "MPLS and the Evolving Internet Architecture", IEEE Comm. Mag., Dec. 1999, pp. 38 – 41

[7] A. Viswanathan, N. Feldman, Z. Wang, R. Callon, "Evolution of Multiprotocol Label Switching", IEEE Comm. Mag., May 1998, pp. 165 – 173

[8] T. M. Chen, T. H. Oh, "Reliable Services in MPLS", IEEE Comm. Mag., Dec. 1999, pp. 58 – 62

[9] K. Habara, T. Matsunaga, K.-i. Yukimatsu, "Large-Scale WDM Star-Based Photonic ATM Switches", J. Lightwave Tech., vol. 16, no. 12, Dec. 1998, pp. 2191 – 2201

[10] J. S. Turner, "Terabit burst switching", J. of High Speed Networks, vol. 8, no. 1, 1999, pp. 3 – 16

[11] C. M. Qiao, M. S.Yoo, "Optical burst switching (OBS) - a new paradigm for an optical Internet", Journal of High Speed Networks, vol.8, no. 1, 1999, pp. 69 – 84

[12] S. L. Danielsen, B. Mikkelsen, C. Joergensen, T. Durhuus, K. E. Stubkjaer, "WDM packet switch architecture and analysis of the influence of tuneable wavelength converters on the performance", J. Lightwave Tech., vol. 15, no. 2, Feb. 1997, pp. 219 – 227

[13] S. Baroni, P. Bayvel, "Wavelength requirements in arbitrarily connected wavelength-routed optical networks", J. Lightwave Tech., vol. 15, no. 2, Feb. 1997, pp. 242 – 251

[14] X. Jia, "A distributed algorithm of delay bounded multicast routing for multimedia applications in wide area networks", IEEE/ACM Trans on Networking, vol. 6, no. 6, Dec. 1998, pp. 828 – 837.

[15] C.-S. Wu, G.-K. Ma, B.-S. P. Lin, "Extended Baseline Architecture for Nonblocking Photonic Switching", J. Lightwave Tech., vol. 15, no. 5, May 1997, pp. 771 – 778

[16] E. G. Churin, P. Bayvel, "Design of Free-Space WDM Router Based on Holographic Concave Grating", Photon. Technol. Letters, vol. 11, no. 2, Feb. 1999, pp. 221 – 223

[17] V. Mikhailov, P. Bayvel, "Low-crosstalk, loss and PDL grating demultiplexer for application in terabit routed networks", to be published in: Proceedings of OFC2000, 5 – 10 March 2000, Baltimore, MD

Analysis and Dimensioning of a Single-Layer Optical Network Based on a "Switchless" Concept in Relevant Scenarios

S. Binetti *, A. Bosco *, M. Listanti [$], A. Maga *, R. Sabella ^

[+] *CoRiTeL – Consorzio di Ricerca sulle Telecomunicazioni, Roma, Italy; E-mail: bosco@coritel.it; fax +39 6 72597439; tel. +39 6 72597441; Via di Tor Vergata 110, 00133, Rome, Italy.*

[$] *INFOCOM Dept – Università di Roma "La Sapienza", Rome, Italy; E-mail: marco@infocom.ing.uniroma1.it; fax +39 6 48906114; tel. +39 6 44585640; Via Eudossiana 18, 00184, Rome, Italy.*

^ *CoRiTeL – Ericsson Telecomunicazioni, Research & Develop. Division, Rome, Italy; Corresponding author: Roberto Sabella; E_mail: sabella@coritel.it; fax +39 06 72597439; tel. +39 6 72597442; Via Anagnina 203, 00040, Roma, Italy.*

Key words: Switchless, WDM, WDMA/TDMA, network analysis and dimensioning, transport network architecture, wavelength converter.

Abstract: The "switchless" all-optical network is an alternative networking approach, developed in the framework of the ACTS project named SONATA (Switchless Optical Network for Advanced Transport Network), which aims to provide a future single-layer, advanced transport architecture on a national scale. The single-hop, shared-access network employs time and wavelength agility, using fast tunable transmitters and receivers. The dimensioning of this type of network is one of the main tasks for the design of the network, which serves a certain number of customers, connected together by means of Passive Optical Networks (PONs). This paper reports an analytical model which allows network dimensioning, according to relevant design parameters: number of customers per PON, number of PONs, offered traffic per single user (residential or business), number of total wavelength converters, and the required system performance expressed in terms of blocking probability.

1. INTRODUCTION

The "switchless" concept was born as an alternative transport network architecture; it has been proposed with the aims of avoiding the need of large and fast switching nodes, simplifying network structure and layer architecture within the network [1], [2]. The proposal consists of a single-layer network platform for end-to-end optical connections, able to serve a large number of terminals, distributed on a wide geographical area, and to handle residential and business traffic. This network model offers a multiple-access, based on a single node providing only passive routing functions [3] - [7]: it is realised by a WDMA/TDMA scheme [1], [2]. On this basis, a terminal is connected to the Passive Wavelength Routing Node (PWRN) by means of a set of wavelengths, each one carrying a TDM flow. In *Figure 1*, it is shown the system scheme.

The structure is formed by N_p PONs, N_t fast tunable terminals (belonging to each PON), N_d re-circulating lines (or "dummy ports"), and N_c lines dedicated to signalling. In this way, the total number of PWRN ports is $N=N_p+N_d+N_c$. In general, cascade optical amplifiers are required, and, in large-split "HyperPONs", a noise gating functional block has to be introduced after those amplifiers for filtering out-band noise [10] [11]. As stated above about the access, a direct connection between two terminals, belonging to the same or to different PONs, is set up by the network control selecting a time-slot on that particular wavelength, which is associated to a specific input PON. This way of working allows us to define the "switchless" concept in the sense that physical switching function is removed from the network, limiting it to the terminals (infact, neither electronic switches/cross-connects [8], nor optical cross-connects [9] are needed). Of course, a centralized network control is required, but architecture simplifications and hardware reductions are obtained; moreover, collisions among calls are prevented by a cyclic assignment of wavelengths at the input of PWRN and by TDMA scheme. In *Table 1*, it is reported an example of wavelength assignment. To achieve a higher degree of flexibility in the use of network capacity and to support traffic that cannot be served directly (PON-to-PON connection), a set of N_d wavelength converter arrays, on re-circulating lines, is introduced (they offer auxiliary capacity to PWRN). In this way, a variable number of wavelengths, depending on traffic requirements, can be dynamically allocated between any pair of terminals: if a direct connection is not available, a call will be routed from a source PON towards the destination PON via a proper wavelength conversion, by using a free time-slot on a wavelength connecting the source PON with a "dummy port", and the *same* time-slot on the wavelength connecting that "dummy port" to the destination PON.

In §1, it has been presented the "switchless" network scheme; in §2, it will be presented the analytical model, with network dimensioning considering a telephone traffic uniformly distributed (§2.1), and a telephone traffic with unbalanced distribution and limitation of wavelength converters number (§2.2); in §3, conclusions will be presented.

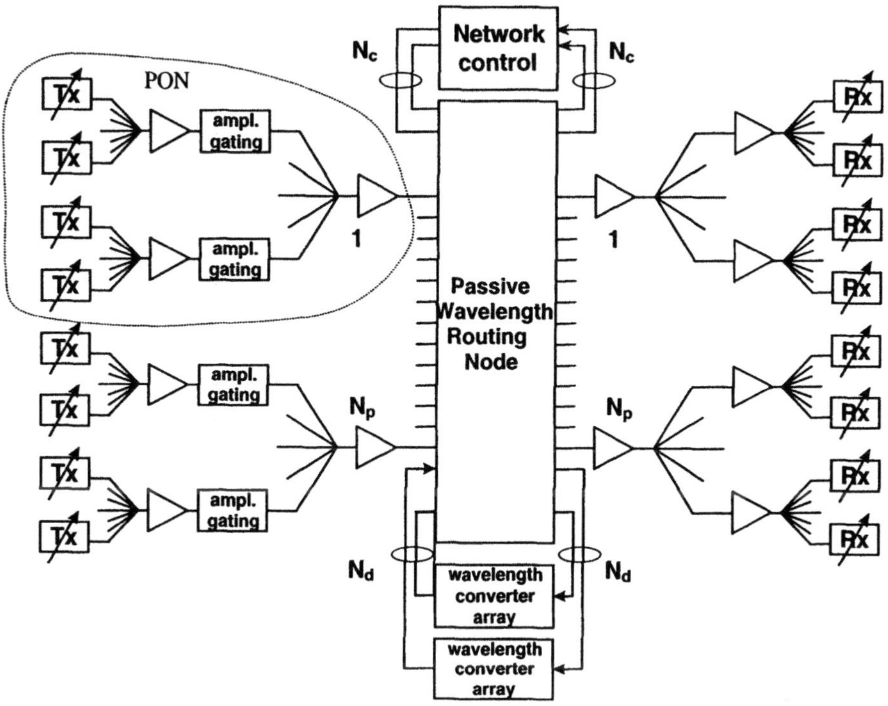

Figure 1. Complete "switchless" network structure

Table 1. Example of wavelengths conflict free allocation in 4x4 PWRN with Np=4 and Nd=2

		N_p ports				N_d ports	
		1	2	3	4	5	6
N_p ports	1	1	2	3	4	5	6
	2	6	1	2	3	4	5
	3	5	6	1	2	3	4
	4	4	5	6	1	2	3
N_d	5	3	4	5	6	1	2
	6	2	3	4	5	6	1

2. MODELLING FOR NETWORK DIMENSIONING

To assess the feasibility of the "switchless" network concept, it is necessary to investigate how the network should be dimensioned in terms of available resources, considering relevant network scenarios and realistic traffic distributions. In the framework of the ACTS-SONATA project, three relevant cases have been studied: small, medium, and large scale networks, referring to residential and business customers. In the analysis, it has been considered telephone traffic, both with uniform and unbalanced distribution. The choice of telephone traffic allows both the utilisation of Poisson traffic model and the achievement of meaningful results, so highlighting the main system properties.

2.1 Analytical model - uniform traffic distribution

We assume to have uniform traffic distribution between input and output PONs. In order to investigate the network performance, the blocking probability has been evaluated both as a function of network resources (number of PONs, number of "dummy ports", number of wavelength converters per "dummy port") and network requests (number of users). In this sense, the network dimensioning can be accomplished by evaluating network parameters that guarantee a blocking probability less than a fixed value, which represents a network quality parameter (e.g. 10^{-5}). An exact computation of the blocking probability would require a solution of an N_p-dimensional Markov chain, with a states number equal to T^{N_p}: such a model becomes too complex already with few PONs and a little time-slots number (T).

To overcome this problem, it has been analysed a *single traffic relation* between the i-th input PON and the j-th output PON, and it has been evaluated the corresponding blocking probability taking into account traffic effects, related to the other traffic relations. According to uniform traffic hypothesis, this probability is also the blocking probability associated with the whole system. The logical model is shown in *Figure 2.* In *Table 2*, the notations used are defined.

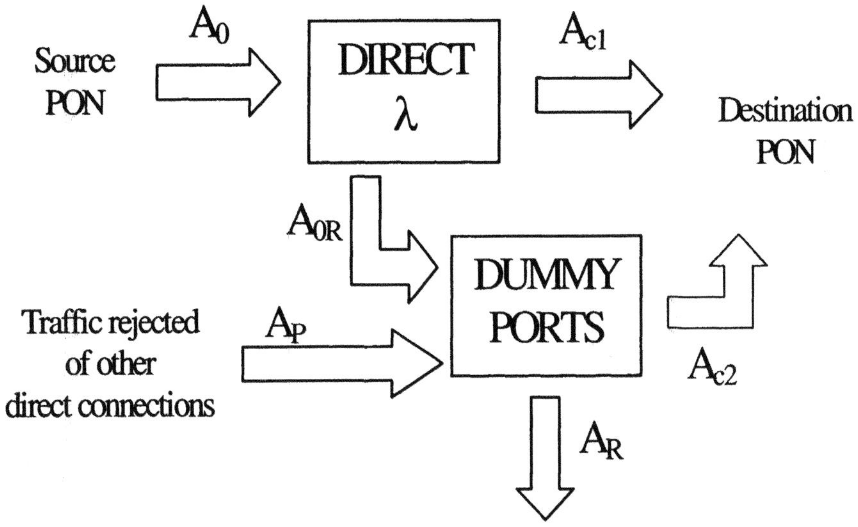

Figure 2. Scheme of the analytical model

In order to analyse the model sketched in *Figure* 2, the following simplifying hypotheses have been done:

- wavelengths connecting the source PON to the "dummy ports" are considered to be blocking free. This means that input PONs have a dedicated wavelength converter per "dummy port", in order to establish connection input PON-output PON through a suitable wavelength conversion at the output of a "dummy port". This assumption allows the evaluation of the blocking probability due to the "dummy ports", as exclusively depending on the saturation of time-slots of a specific wavelength connecting the "dummy ports" with the destination PONs.
- traffic refused by direct wavelengths and offered to the "dummy ports" ($A_{o,R}$, A_P) is represented by only its first two moments (the mean and the variance); this assumption allows the application of the Equivalent Random Traffic (ERT) method [12] for the analysis of the blocking probability of the wavelength outgoing from the "dummy ports".

Four steps support the analysis:

1. evaluation of the loss probability (π_1) experienced on the direct wavelengths (PON-PON): this is the probability that a generic call can not be handled by direct service, and it must go towards "dummy ports";
2. evaluation of the traffic refused by "dummy ports" and calculation of the relative blocking probability (π_2);

3. calculation of the total loss probability (π): this is the probability that a generic call cannot be accommodated anyway.

Table 2. Notations for the network parameters used

Parameter	Definition
A_o	Average traffic intensity offered for single relation input/output;
$A_{o,tot}$	Average traffic intensity offered globally ($A_{o,tot}=N_p*A_o$);
A_{c1}	Average traffic carried by a direct wavelength between an input and an output PON;
$A_{o,R}$	Average traffic rejected for single relation input/output and offered to the "dummy ports";
A_P	Average traffic rejected by all direct wavelengths and offered to the "dummy ports" ($A_P=(N_p-1)*A_{O,R}$);
A_{c2}	Average traffic carried globally by the "dummy ports";
A_R	Average traffic rejected by the system;

On the basis of what said until now, we have three different blocking probabilities: the first at the input of direct service (input of router ports), the second at the input of "dummy ports", the third at the output of "dummy ports". In order to calculate the first loss probability, considering that direct wavelengths connecting the source and the destination PONs are considered as a group of T servers with at the input a Poisson traffic of average value A_o, it has been applied the Erlang-B analytical model; thanks to that, parameters as π_1 (1), $A_{o,R}$ (2), standard deviation $\sigma^2_{o,R}$ (3), and the total traffic refused (4) by direct service can be calculated:

$$\pi_1 = B(A_o, T) = \frac{A_o^T / T!}{\sum_{j=o}^{T} A_o^j / j!} \tag{1}$$

$$A_{o,R} = A_o \cdot \pi_1 \tag{2}$$

$$\sigma_{o,R}^2 = A_{o,R} \cdot (1 - A_{o,R} + \frac{A_o}{T + 1 - A_o + A_{o,R}}) \tag{3}$$

$$A_{o,dummy} = A_{o,R} + A_P = A_{o,R} + (N_P - 1) \cdot A_{o,R} = N_P \cdot A_{o,R} \tag{4}$$

Similarly, in order to calculate the blocking probability at the input of "dummy ports", traffic rejected by "dummy ports" must be calculated. This traffic is known to be peaked: so, it cannot be applied the Erlang model. However, according to the ERT method, this traffic can be thought as the overflow traffic of a group of N^* servers (called preamble system) loaded by Poisson traffic of average value A^*. These equivalent parameters can be calculated, solving the following two-equation system (5):

$$\begin{cases} A_{o,dummy} = A^* \cdot B(A^*, N^*) \\ V_{o,dummy} = A_{o,dummy} \cdot \left(1 - A_{o,dummy} + \dfrac{A^*}{N^* + 1 - A^* + A_{o,dummy}} \right) \end{cases} \tag{5}$$

Once A^* and N^* are known, the mean value of the traffic overflowing the "dummy port" system (6) and the blocking probability at the input of "dummy ports" (7) are given by:

$$A_R = A^* \cdot B\left[A^*, \left(N^* + N_d \cdot T\right)\right] \tag{6}$$

$$\pi_{in} = {}^{A_R}\!\big/_{A_{o,dummy}} \tag{7}$$

Now, to complete the analysis and to find the total blocking probability, we must take into account conflicts at the output of the "dummy ports": to do this, "dummy ports" have been considered as a cascade of two gates, in which offered traffic can be rejected by the second block. In this case the evaluation of the second loss probability is given by:

$$\pi_2 = 1 - (1 - \pi_{in}) \cdot (1 - \pi_{out}) \tag{8}$$

where:

- $(1-\pi_{in})$ is the probability that there is at least one slot for access to re-circulating line, considering traffic rejected by direct connections;
- $(1-\pi_{out})$ is the probability that there is at least one slot to leave the re-circulating line, considering only the traffic that has passed the input block of "dummy ports".

The blocking probability at the output of the "dummy ports" is calculated by Frederick's model.

To be noticed that this model is not rigorous: it doesn't take into account the coincidence of time-slots at the input and output of "dummy ports". The total blocking probability is given by $\pi = \pi_1 * \pi_2$.

2.1.1 Network dimensioning results for uniform traffic distribution

To investigate system performance, we consider three different scenarios (large, medium, small region), corresponding to different sizes of a region served by the system, and two user classes (residential and business); the data used for the study are reported in *Table 3.* A single business user presents a traffic intensity of 0.15 Erl, while a residential user presents a traffic intensity of 0.05 Erl; the average call holding time is supposed to be equal for both classes. The scenarios introduced are analysed considering a) full residential (100% of residential users); b) mixed (78% of residential users and 22% of business users); c) full business (100% of business users). Finally, T = 1000 is the maximum number of time-slots of the TDM flow dedicated to the telephone traffic on each wavelength line accessing the PWRN.

Table 3. Network and traffic scenarios for system performance analysis

Network scenarios		**Traffic scenarios**		
Type	**Number of users**	**Residential**	**Mixed**	**Business**
Small region	15000	750 Erl	1080 Erl	2250 Erl
Medium region	150000	7500 Erl	10800 Erl	22500 Erl
Large region	2000000	100000 Erl	144000 Erl	300000 Erl

In *Figure 3*, for instance, are reported the values of the PWRN port capacity, $C_s(N_p)$, expressed in number of time-slots vs. the number of PONs, for the different considered scenarios. To be noticed that $C_s(N_p)$ is defined as the minimum between 1000 time-slots and the minimum value to guarantee the 1th loss probability (only for direct service) less than 0.1. From this figure, we can see that increasing the number of PONs, $C_s(N_p)$ decreases because the total offered traffic per PWRN port decreases. Moreover, it is intuitive to understand that, for a given loss probability and PWRN capacity, increasing the number of PONs, the number of "dummy ports" (or, equivalently, the auxiliary capacity on re-circulating lines) decreases, because calls, on average, can be handled through direct connections

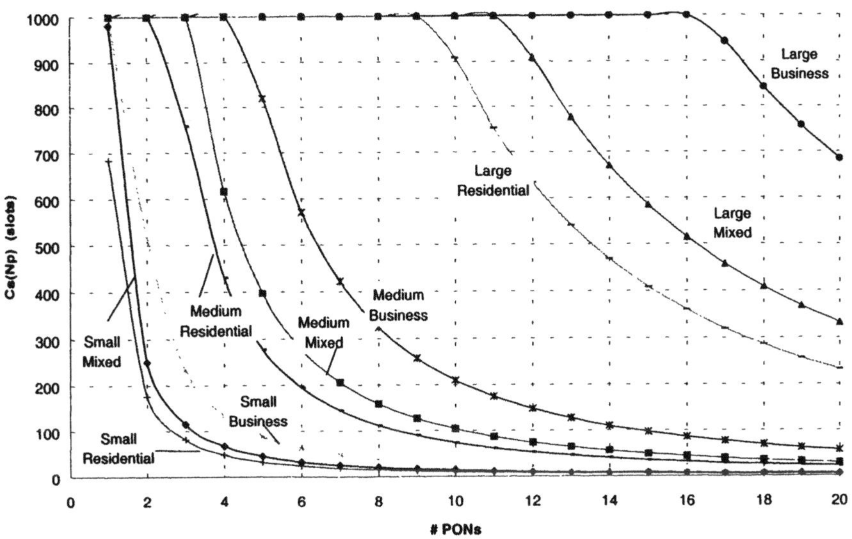

Figure 3. Number of slots vs. number of PONs in considered system scenarios

A parameter that can be chosen for representing the system cost is the total number of ports (N_t) of the PWRN. In *Figure 4*, it is sketched the total number of ports vs. the number of PONs (N_d):

- for low N_d values, the PWRN size decreases as the number of "dummy ports" increases;
- for higher values of the number of "dummy ports", PWRN size linearly increases with N_d.

The threshold between the two zones is the best choice for system configuration.

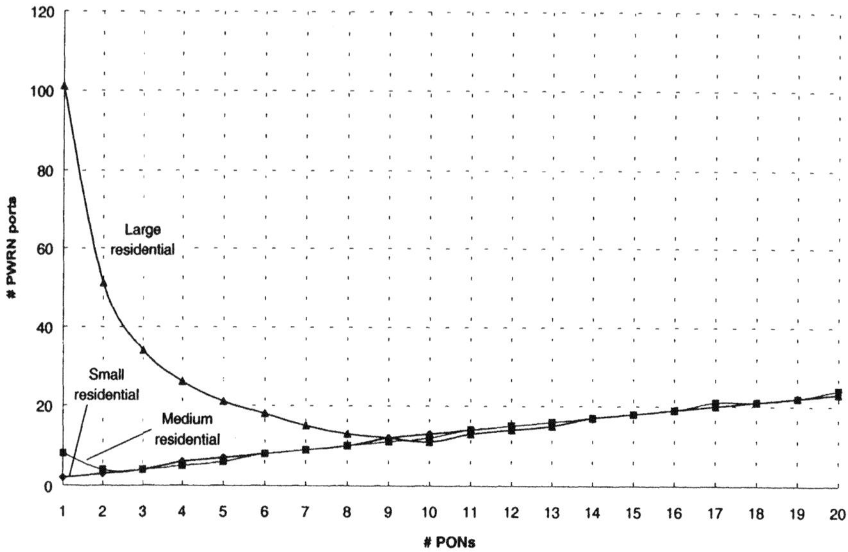

Figure 4. Total number of PWRN ports (Nt) vs. number of PONs (Nd) in case of residential traffic scenario (constraint: $\pi<10^{-5}$)

In order to understand the real potentiality of the "switchless" network, in *Figure 5*, the number of users that can be supported by a single PWRN vs. the number of "dummy ports" is shown, considering the constraint $\pi < 10^{-5}$. The data reported in this figure suggest that this kind of system could be used to serve a wide area, comparable with a medium-large country, with a lot of users.

2.1.2 Comparison between analytical and simulation models

In order to test the reliability of the analytical model, we have realised a simulative model, obtaining, for the same relevant cases, several results; such results have been compared with those arising by analytical model. The comparisons have shown that the difference among results obtained by analytical model (using Wilkinson and Fredericks model, considering the block at the output of "dummy ports"), and those obtained by simulation model is negligible. Moreover, it has been observed that the contribution of the block point at the output of "dummy ports" to the calculation of global blocking probability is negligible. On this basis, all the other results have been obtained neglecting the contribution of the block at the output of "dummy ports".

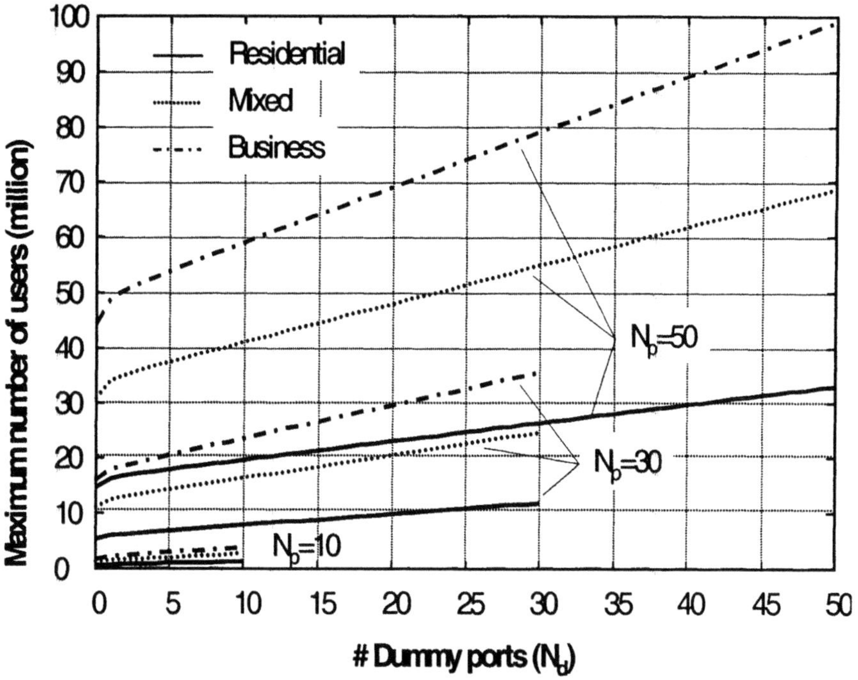

Figure 5. Maximum number of users vs. the number of dummy ports (Nd) for $\pi<10^{-5}$

2.2 Analytical model - unbalanced traffic distribution

Until now, it has been considered a uniform traffic distribution among PONs. Now, we remove this limitation and consider the more realistic case of an unbalanced traffic distribution: some PONs offer more traffic to the PWRN than others do. We assume that the total offered traffic by all PONs follows a gaussian distribution: there will be a single PON which offers, on average, more traffic than others PONs, with a certain standard deviation. Of course, on time, PON, which offers more traffic, will not be always the same, but we imagine a situation variable in time. Traffic type is still supposed only telephone. The analytical model is the same described in §2.1. There are only two differences: firstly, traffic parameters must be expressed as a sum of *different* terms:

$$A_{o,tot} = \sum_{i=1}^{N} A_o(i) \neq N * A_o$$

The second difference is that the requested number of "dummy ports" depends on PON activity: the PON which offers more traffic uses more "dummy ports" or wavelength converters; the PON which offers less traffic uses less "dummy ports".

2.2.1 Network dimensioning results for unbalanced traffic distribution

We have considered two possibility choices for network dimensioning: the first in which we choose a number of "dummy ports" equal to the maximum value requested by that PON which offers more traffic, and the number of wavelength converters is equal to the number of PONs. In the second we maintain constant the total available resources, increasing the number of "dummy ports", but reducing the number of wavelengths per "dummy port".

The major results obtained with the first network dimensioning type are shown below.

In *Figure 6*, assuming 10 input PONs belonging to the PWRN ports and a standard deviation equal to 2, it is shown the blocking probability vs. the number of "dummy ports", for each PON; it is also shown, as comparison, the curve relative to the mean value of loss probability.

It can be seen that the PON that offers more traffic (in this graphic it is represented by curve labelled with the number 5) requires more "dummy ports" (offering more traffic to the PWRN, it needs more wavelength converters, that is, more auxiliary capacity), for a given blocking probability.

In *Figure 7*, it is shown the number of wavelength converters vs. the number of PONs, varying standard deviation; it is included also a curve for uniform traffic distribution. It can be seen that increasing the standard deviation value, system behaviour is close to system behaviour in uniform traffic case. Moreover, increasing number of PONs, number of wavelength converters decreases because, for a given total offered traffic at input of the system, traffic offered per each relation input PON i/output PON j decreases.

The major results obtained with the second dimensioning choice will be shown below.

It is worth to notice that in this case, considering that now the number of wavelength converters per "dummy port" is limited to a number less than the number of input PONs, we must use tunable wavelength converters in the re-circulating lines. Besides, considering that the number of these devices are limited, another block point must be taken into account: infact, a call cannot be handled if there isn't a free wavelength converter in the re-circulating lines. Also this blocking probability is calculated by the Fredericks method.

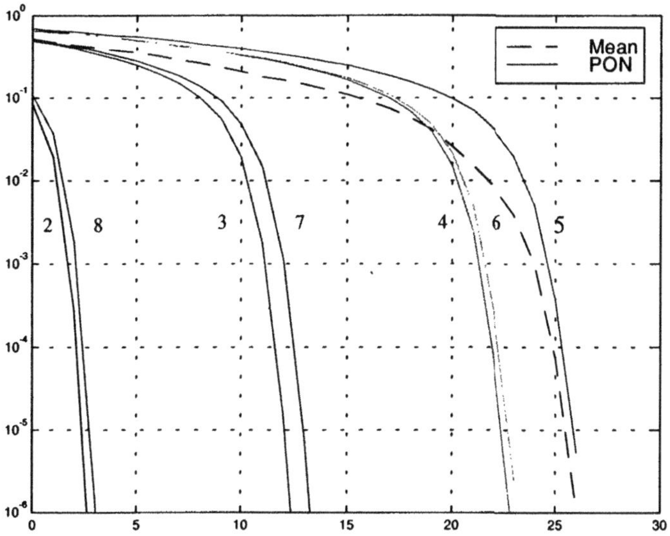

Figure 6. Blocking probability (ordinate) vs. the number of "dummy ports" (abscissa), for 10 PONs (each color is relative to a PON)

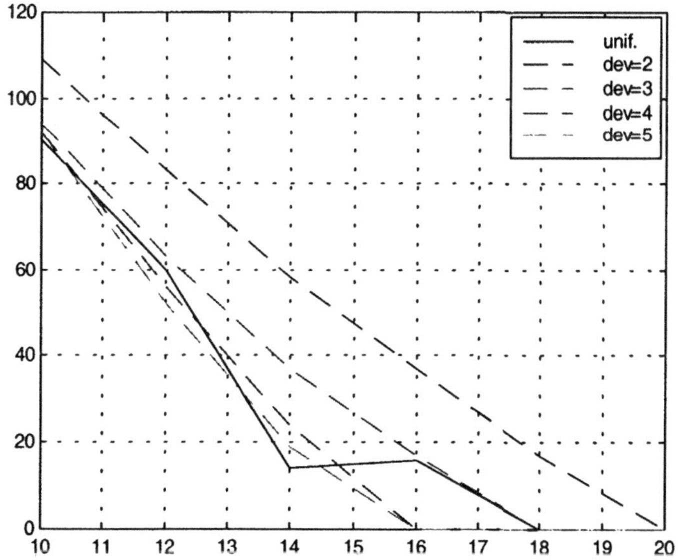

Figure 7. Number of wavelength converters (ordinate) vs. the number of PONs (abscissa), varying standard deviation

In *Figure 8*, assuming a uniform traffic distribution among PONs, it is shown the blocking probability vs. the total number of wavelength converters, varying the number of wavelength converters per "dummy port" (that is, the number of wavelength converters effectively needed). It can see that, diminishing the number of wavelength converters per "dummy port", their total number decreases; however, this variation is not remarkable (for a blocking probability of 10^{-5}, the number of wavelength converter change from 93, with WC=10, to 70, with WC=2).

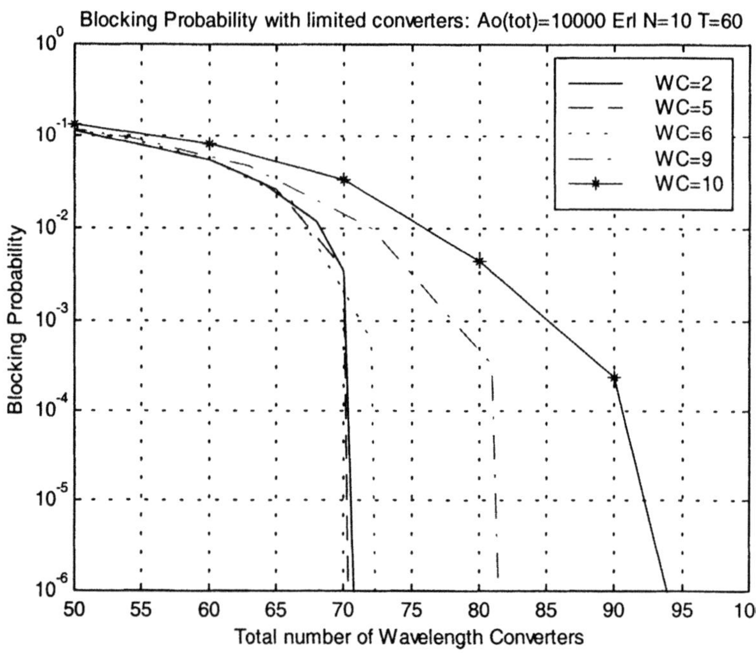

Figure 8. Blocking probability vs. the total number of wavelength converters, varying the number of wavelength converters (WC) per "dummy port", assuming a uniform traffic distribution among PONs

Different result is obtained considering an unbalanced traffic distribution among PONs, as shown in *Figure 9*: using five wavelength converters per "dummy port", the reduction in the total number of wavelength converters is very remarkable (from about 250 to 130 wavelength converters); it justify the higher cost and higher control complexity because of the use of tunable converters.

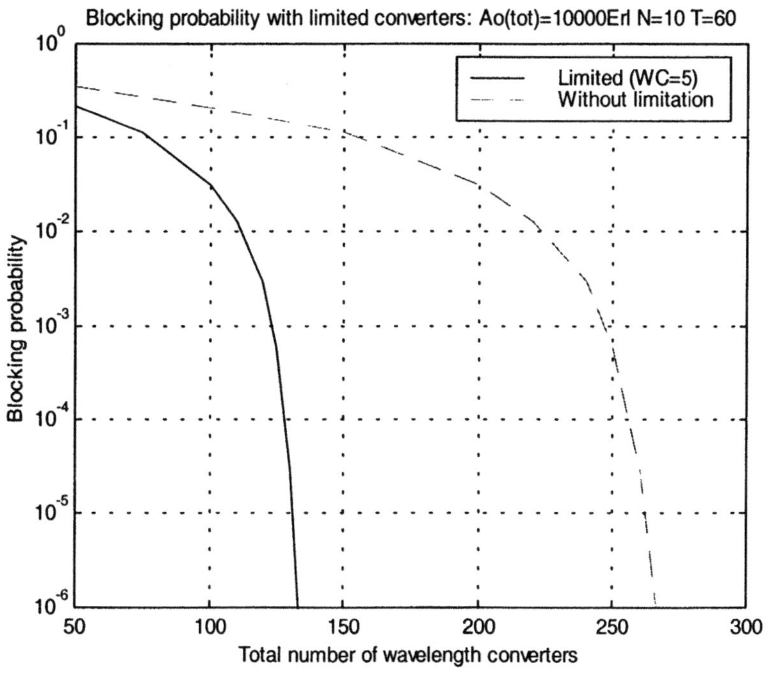

Figure 9. Blocking probability vs. the total number of wavelength converters, varying the number of wavelength converters per "dummy port", assuming an unbalanced traffic distribution among PONs

2.3 Comparison between analytical and simulation models

The validity of these models, under the hypotheses made, has been verified. The comparison, in the case of no limitations on wavelength converters, has shown that analytical and simulation models are very close; they are not coincident because, in simulation model, time –slots coincidence at the input and output of "dummy ports" have not been taken into account.

In case of limitation of wavelength converter per "dummy port", the validity of the models, and the hypotheses made have been verified through another simulation model. Also in this case, the comparison has shown that analytical and simulation models have a very close behaviour.

3. CONCLUSIONS

In the framework of ACTS-SONATA project, it has been developed an analytical model that allows "switchless" network dimensioning, in the case of Poisson and Gaussian traffic, considering different user classes. These models have been demonstrated to be in full agreement with simulation models, also developed within the same project. The number of customers in a PON, the number of PONs, the number of re-circulating lines (i.e. the number of wavelength conversion blocks or "dummy ports"), and the total number of wavelength converters have been considered as design parameters. As results, the analyses reported for some relevant cases showed that is possible to assess the feasibility of a "switchless" network, since there are no constraints in realising networks covering even for large-scale areas.

On the basis of a gaussian traffic distribution, it has been seen that the PON which offers more traffic requires more "dummy ports"; moreover, the more unbalanced is the traffic offered, the higher is the number of "dummy ports" for a given blocking probability.

About network dimensioning, limiting wavelength converters number per "dummy port" and increasing these, we have found that advantages, in terms of reduction of total wavelength converter number, can be obtained in both cases of uniform and unbalanced traffic distribution: but, while in the first case the reduction of total converters is little, in presence of an unbalanced traffic distribution the advantages is higher, justifying the increment in cost and control complexity due to the use of tunable wavelength converters, on the re-circulating lines.

4. BIBLIOGRAPHY

[1] A. Hill, A. Carter, S. Armitage, J. Shabeer, M. Harmon, P. Rose, "A scalable and switchless network structure, employing 32x32 free-space grating multiplexer", IEEE Photonics Technol. Lett., vol.8, pp. 569-571, 1996.

[2] N. P. Caponio, A. M. Hill, R. Sabella, "Switchless Optical Network for Advanced Transport Architecture" LEOS'98, Paper WZ4, Orlando, Dec.1998, pp.358-359.

[3] T. Durhuus, et al., "All-optical wavelength conversion by semiconductor optical amplifiers", IEEE Journal of Lightwave Technology, vol. 14, no. 6, pp. 942-954, 1996.

[4] C. Joergensen, et al., "Wavelength conversion by optimized monolithic integrated Mach-Zehnder interferometer ", IEEE Photonics Technology Letters, vol. 8, no. 4, pp. 521-524, 1996.

[5] W. Idler, et al., "10 Gb/s wavelength conversion with integrated multi-quantum well based 3 ports Mach-Zehnder interferometer", IEEE Photonics Technology Letters, vol. 8, no. 9, pp.1163-1165, 1996.

[6] J. Leuthold, et al., "All optical 2x2 switches with 20 dB extinction ratios", Electronics Letters, vol. 32, no. 24, November, pp. 2235-2236, 1996.

[7] F. Ratovelomanana, et al., "Improvement of regeneration capabilities of an all optical wavelength converter, based on a Mach-Zehnder interferometer, by means of a phase-shifter section", IEE Electronics Letters, vol. 33, no. 19, pp. 1629-1630, 1997.
[8] K. Sato, Advances in Transport Network Technologies, Norwood, MA: Artech House, 1996.
[9] R. Sabella, P. Lugli, High Speed Optical Communications, chap. 12, Kluwer Academic Publishers, 1999.
[10] A.M. Hill, "WDM Networking in the Access Network and Beyond", ECOC'95 Optical Networking Forum Workshop, Paper S1.4, 1995.
[11] N. P. Caponio, A.M. Hill, F. Neri, R. Sabella, "Single-layer optical platform based on WDM/TDM multiple access for large-scale "switchless" networks", in press on European Transactions on Telecommunications, Special Issue on WDM Networks.
[12] J. Y. Hui, Switching and Traffic Theory for Integrated Broadband Networks, Kluwer Academy Press, 1990.

Stefano Binetti was born in Rome, Italy, in 1973. He received the degree in electronic engineering from University of Rome "Tor Vergata" in 1998. His research interests are in high-speed optical communications systems and WDM optical networks. In March 1998 he won a research fellowship and then, he joined CoRiTeL (Rome, Italy); he worked in the ACTS Project "SONATA". Now, he is with Pirelli Optical Systems where he is involved in a research & development activity on optical networks. He is also following the standardisation activity on this topic.

Alessandro Bosco was born in Rome, Italy, on May 1973. He received the degree in telecommunication engineering from University of Rome "Tor Vergata" in 1999. His research interests are in High Speed Optical Systems, WDM optical networks and Physical Layer Modeling. In September 1999, he won a research fellowship and then, since September 1999, he joined CoRiTeL (Rome, Italy). Now he is working in the ACTS 351 Project named "SONATA".

Marco Listanti received his Dr. Eng. degree in electronics engineering from the University "La Sapienza" of Roma in 1980. He joined the Fondazione Ugo Bordoni in 1981, where has been leader of the group "TLC network architecture" until 1991. In November 1991 joined the INFOCOM Dept. of the University of Roma "La Sapienza", where he is Associate Professor in Switching Systems. Since 1994, he also collaborates with the Electronic Department of the University of Rome "Tor Vergata" where he holds courses in Telecommunication networks. He participated at several international research projects sponsored by EEC and ESA and is author of several papers published on the most important technical journals and conferences in the area of telecommunication networks. His current research interests focus on traffic control in IP networks and on the evolution of techniques for optical networking. Mr. Listanti has been representative of

Italian PTT administration in international standardisation organisations (ITU, ETSI). Mr. Listanti is also a member of IEEE Communications and Computer Societies.

Roberto Sabella received the degree in electronics engineering (Laurea in Ingegneria Elettronica) from the University "La Sapienza" of Roma, Italy, in 1987. He then joined Ericsson Telecomunicazioni, Rome, Italy, where he was a Hardware Designer for about one-and-a-half years, and then as a Researcher on advanced fiber-optic communications. His research interests are in high-speed optical communication systems and optical networks. Since 1991, he led a group working on the modeling and simulation of advanced communication systems. He holds two patents on optical cross-connects, is co-author of a book on high-speed optical communications and author of more than 70 papers on scientific/technical journals and international conferences. Since May 1997, he has been with CoRiTel consortium as research technical co-ordinator. He has been Lecturer (professore a contratto) at the University of Rome "Tor Vergata" and at the Polytechnic of Bari, Italy. Mr. Sabella is member of the IEEE/LEOS Technical Committee on Optical Networks and Systems, and member of the editorial board of the new international journal Photonic Network Communications (New York: Kluwer Academic). For the same journal he has operated as guest editor for the special issues on WDM transport networks. Moreover, he has been invited to be one of the guest editors of a special issue on optical networks, for the journal Computer Networks (Elsevier).

Optical CDMA Networks with Bipolar Codes

A. Gameiro
Univ. Aveiro / Instituto Telec., Campus Universitário, 3810-193 Aveiro, Portugal

Key words: Optical CDMA, Networks, Spread-Spectrum

Abstract: In this communication we consider a dual wavelength scheme that allows the use of bipolar codes in optical code division multiple access networks. The performance of the system is derived for asynchronous and asynchronous users, and the results show that significant throughput improvements are achievable relatively to the conventional schemes relying on unipolar codes.

1. INTRODUCTION

Optical fiber networking will be undoubtedly one of the key components to meet the growing demands of the information society. At the core of the development is the huge bandwidth of a single mode fiber that can support up to several Tbit/s transmission capacity. Due to the speed limitation of signal processing in the electrical domain, optical multiplexing techniques have to be employed to exploit the full transmission capacity.

Presently there is some consensus on the use of wavelength division multiplexing [WDM] for more efficient utilization of the bandwidth of existing optical fiber links, upgrading their capacity in a cost effective manner [1]. The implementation of dense WDM requires however high precision wavelength stabilization techniques as well as the optical frequency standard. An alternative to WDM is code division multiple access (CDMA), or code division multiplexing (CDM), where the signals from different terminals coexist either in time or frequency, and are discriminated through an unique code or signature attributed to each user. CDMA relies on spread spectrum transmission techniques, and has gained a wide acceptance in radio communications: it is used in several commercial systems, and has already been specified as the technology for third generation cellular systems. Code division multiplexing offers some advantages over the more traditional techniques based on time or frequency discrimination, namely:

- It allows uncoordinated access to the common transmission medium;
- It presents a soft performance degradation as a function of the number of users, which may be exploited when considering applications requiring different quality of services;
- It presents some inherent security.

These advantages result in a more flexible network requiring a less sophisticated management when compared to FDM or TDM. Furthermore in optical fiber networking CDM can be elegantly overlaid on the existing WDM networks providing an enhanced versatility and helping to solve the future wavelength resource problems [2]. It is therefore expectable that similarly to what happened in radio communications, the CDM techniques will be a key component of optical multiplexing in the medium term, and in fact considerable research on this topic has been carried out in the last years. The work on optical CDMA (OCDMA) has been focused towards two different categories: 1) coherent systems using phase-shifted code sequences [3-7] and 2) non-coherent systems adopting intensity modulated sequences [2, 8-10]. Non-coherent systems have significant advantages in terms of the complexity required to build the transmitters and receivers, but because they are based on intensity modulation, the signature codes are restricted to unipolar (0,1) sequences. This non-negative nature of the signals, prevents the use of the well studied bipolar codes commonly used in radio communications and provided the impetus for the development of unipolar pseudo-orthogonal codes [3, 5,11-12]. The number of available codes is however rather low and they are clearly inferior in terms of auto-correlation or cross-correlation properties to the well designed bipolar codes such as Gold or Kasami sequences, thus putting limits on the maximum number of simultaneous users. To allow the use of bipolar codes and thus take advantage of the extensive material available on the design of bipolar sequences with good properties, one solution is to go to optical coherent communications. The option for coherent communications represent however a significant increase in the complexity of transceivers: the coders / decoders must employ phase shifters to reverse polarity, and receivers require phase feedback control loops which are difficult to engineer. Motivated by these facts we consider in this communication a solution of intermediate complexity: the coders / decoders are based on phase-shifted optical sequences enabling the use of bipolar codes, but information corresponding to logical ones and logical zeros is sent in different wavelengths which are detected non-coherently. This constitutes basically a non-coherent FSK system. The communication is outlined as follows. In section 2 we present the architecture of the system we will be dealing with and describe the transceivers for the non-coherent system suitable to operate with bipolar codes. In sections 3 we derive the performance of the proposal for both cases of synchronous and asynchronous networks, and based on these results present in section 4 numerical examples and comparisons against the corresponding systems based on unipolar codes. Finally in section 5 we outline the main conclusions of this work.

2. SYSTEM MODEL

The general architecture of the network we are dealing with is shown in Fig. 1. There are N transmitter / receiver pairs (users), and each user is assigned a unique signature or code sequence to assess the common medium. The block diagram of the transmitter used at each station is shown in Fig. 2. Two pulsed laser diodes (LD), one operating at wavelength λ_0 and the other at wavelength λ_1 are used. Each LD outputs a train of pulses with duration τ, and period T, where $1/T$ is the data bit rate. The data sequence modulates the intensity of the first LD, and the complementary data the intensity of the second. With a logical 1 in the incoming data we get a pulse of duration τ at the output of the upper modulator and no signal at the output of the lower,

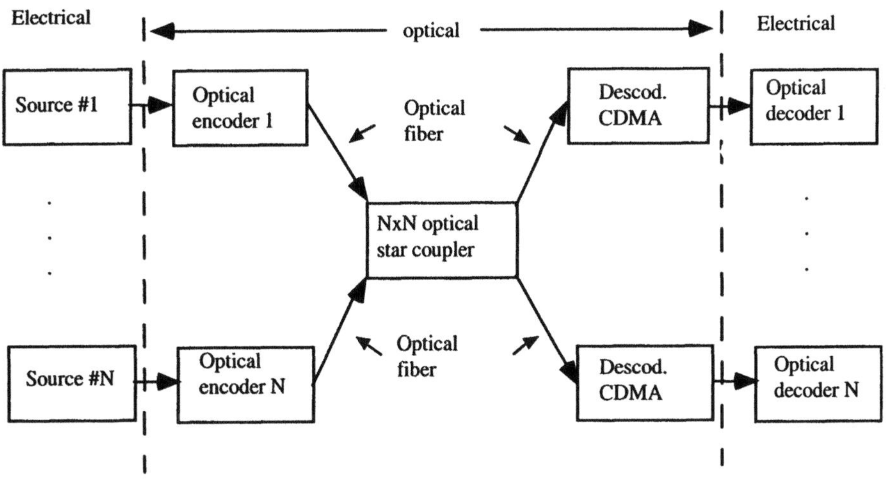

Figure 1. Optical CDMA system in a star configuration.

while for a logical 0 the situation is reversed. The transmitter encoder may be implemented using the fiber optic transversal filters of Fig. 3, which have been experimentally demonstrated in [2] and [9], although other ladder network structures are also feasible [13]. The phase shifts of the various taps are tuned to be 0 or π, which allows one to generate with N taps any PSK code of length N.

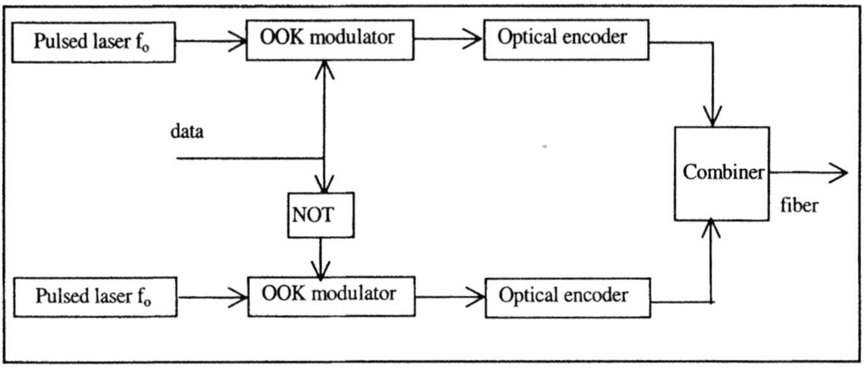

Figure 2. Block diagram of transmitter

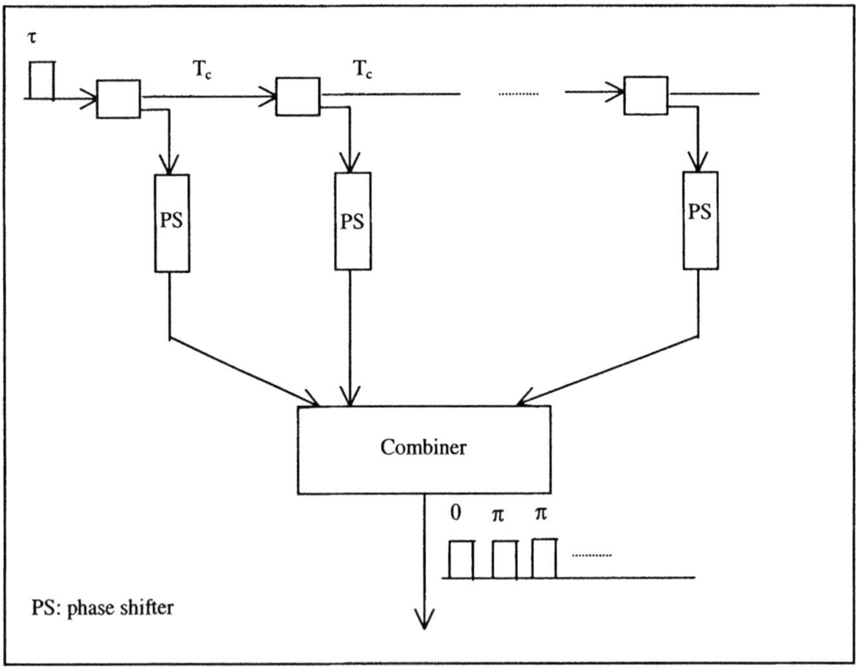

Figure 3. General structure of all optical encoder/decoder.

The optical transmitted signal can then be written as

$$y(t) = \sqrt{2P_T}\left(\sum_{k=-\infty}^{\infty} a_k c(t-kT)e^{j2\pi f_0 t} + \sum_{k=-\infty}^{\infty}(1-a_k)c(t-kT)e^{j2\pi f_1 t} \right) \tag{1}$$

where $\{a_k\}$ is the unipolar data sequence, $f_0 = c/\lambda_0$, $f_1 = c/\lambda_1$ and P_T the peak optical power.

The code waveform $c(t)$ is given by:

$$c(t) = \sum_{j=0}^{N-1} c_j p_\tau (t - jT_c) \tag{2}$$

where $p_\tau(t)$ is a rectangular pulse of duration τ, and $\{c_j\}$ the bipolar (± 1) code sequence.

The structure of the receiver is shown in Fig. 4. The optical received signal is splitted in two arms. An optical filter tuned to the optical frequency f_o, extracts the first component in the right hand side of (1), while the filter tuned to f_1, recovers the second term. An optical decoder matched to the encoder used in the transmitter, i.e. with the taps reversed, follows each optical filter. This gives at the sampling points kT, an intensity peak in the upper arm if the data symbol is a logical one and a peak in the lower arm if the data symbol is a logical 0. Each decoder is followed by a photodetector and the difference applied to a decision device according to the principles of conventional non-coherent FSK receivers.

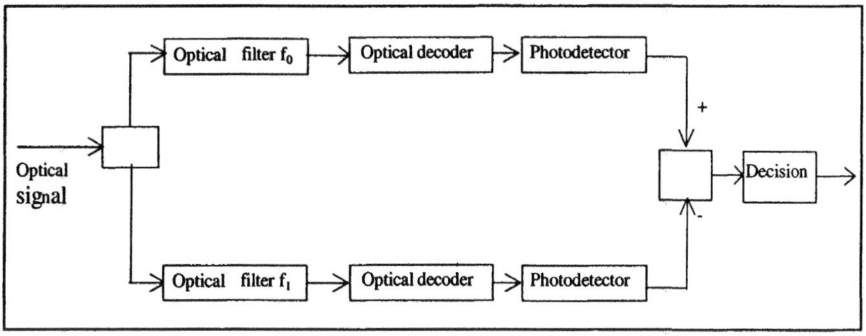

Figure 4. Block diagram of the receiver.

3. SYSTEM ANALYSIS

In the analysis we shall consider the two cases of synchronous and asynchronous users In a synchronized network, there is some centralized control that provides the correct timing to each station so that the signals from the different stations arrive synchronous, i.e. time aligned at the central NxN optical coupler. In an unsynchronized network, there is no timing coordination between the stations and the signals arrive at the coupler with different timing alignments. A synchronized network may achieve a significantly better performance than an unsynchronized one since good codes are much easier to design with this restriction but it entails an extra complexity, and it may often preferable to set the network with uncoordinated users easing the management tasks.

We will consider without loss of generality the receiver for user 0, and once we are interested in the performance limits imposed by the existence of multi-user interference, we will consider ideal components and neglect other sources of disturbance. The derivation assumes thus a high enough signal to noise ratio so that the system is multi-user interference limited.

Considering first the synchronized network, the optical received signal for K simultaneous users is in complex notation

$$r(t)=\sqrt{2P_R}\sum_{i=0}^{K-1}\left(\sum_{k=-\infty}^{\infty}a_k^{(i)}c^{(i)}(t-kT)e^{j(2\pi f_0 t+\phi_i)}\sum_{k=-\infty}^{\infty}(1-a_k^{(i)})c^{(i)}(t-kT)e^{j(2\pi f_1 t+\theta_i)}\right) \tag{3}$$

where it is assumed that we have a perfect power control.

Let us consider that the undesired frequencies are completely eliminated by each of the optical filters and a code length of N. The complex envelope at the output of the matched decoders at time $t=T$ is given by

$$\begin{aligned} r_0(T)&=\alpha\sqrt{P_R}\,(a_0^{(0)}NE_p+\sum_{i=1}^{K-1}a_0^{(i)}e^{j\phi_i}\sum_{j=0}^{N-1}c_j^{(0)}c_j^{(i)}E_p)\\ r_1(T)&=\alpha\sqrt{P_R}\,((1-a_0^{(0)})NE_p+\sum_{i=1}^{K-1}a_0^{(i)}e^{j\theta_i}\sum_{j=0}^{N-1}c_j^{(0)}c_j^{(i)}E_p)\end{aligned} \tag{4}$$

$E_p = \int_{-\infty}^{\infty} p^2(t)dt$, is the energy of the elementary pulse, $\sum_{j=0}^{N-1} c_j^{(0)} c_j^{(i)} = R_{0,i}(0)$ the cross-correlation between the code of users 0 and i, with the sequences aligned, and α accounts for the losses of the splitter and filter.

The variable used for performing the decision about bit 0 is thus

$$v = |r_0(T)|^2 - |r_1(T)|^2 = \alpha^2 P_R E_p^2 \left| a_0^{(0)} N + \sum_{i=1}^{K-1} a_0^{(i)} e^{j\phi_i} R_{0,i}(0) \right|^2 - \alpha^2 P_R E_p^2 \left| (1-a_0^{(0)})N + \sum_{i=1}^{K-1} (1-a_0^{(i)}) e^{j\phi_i} R_{0,i}(0) \right|^2 \quad (5)$$

From (5), we conclude that if the codes are orthogonal ($R_{0,i}(0) =, \forall i \neq 0$), then the system is multi-user interference free, and the decision variable is simply proportional to $b_0^{(0)} = 2a_0^{(0)} - 1$, requiring a threshold comparison against 0. Bipolar orthogonal codes can be constructed using Hadamard matrices through the iterative procedure

$$\mathbf{H}(0) = [1]$$
$$\mathbf{H}(n) = \begin{bmatrix} \mathbf{H}(n-1) & \mathbf{H}(n-1) \\ \mathbf{H}(n-1) & -\mathbf{H}(n-1) \end{bmatrix} \quad (6)$$

leading to a total of 2^n orthogonal codes with length N=2^n, and consequently a synchronous network can support 2^n simultaneous users.

Let us now consider a network with uncoordinated users. The signals from the different stations arrive at the optical coupler with different delays and consequently the optical signal at receiver 0, is

$$r(t) = \sqrt{2P_R} \sum_{i=0}^{K-1} \left(\sum_{k=-\infty}^{\infty} a_k^{(i)} c^{(i)}(t-kT-\tau_i) e^{j(2\pi f_0 t+\phi_i)} \sum_{k=-\infty}^{\infty} (1-a_k^{(i)}) c^{(i)}(t-kT-\tau_i) e^{j(2\pi f_1 t+\theta_i)} \right) \quad (7)$$

where τ_i is the delay of signal i.

The derivation the bit error rate (BER) performance is done in Appendix A. Assuming that the code sequences can be considered as true random binary sequences, one gets

$$p_b = \frac{1}{2} e^{-\frac{3N}{2(K-1)}} \tag{8}$$

Equation (8) is for well designed bipolar codes such as Kasami or Gold, a conservative upper bound, since the auto-correlation and cross-correlation properties of these codes are better than those of pure random sequences. Comparing against the performance obtained with a fully coherent PSK system, (8) represents a degradation slightly higher than 3dB, i.e. the number of simultaneous users is reduced by a factor of 2 due to the non-coherent nature of the detection process.

4. NUMERICAL RESULTS AND COMPARISONS

To illustrate expected performance of the system in an unsynchronized network, we plot in Fig. 5, the bit error rate as a function of the number of users K, for codes lengths 31, 63, 127 and 255, which are the natural length of Gold codes. The results show that for a target of 10^{-9} for the BER the maximum number of simultaneous users is slightly under 10%.

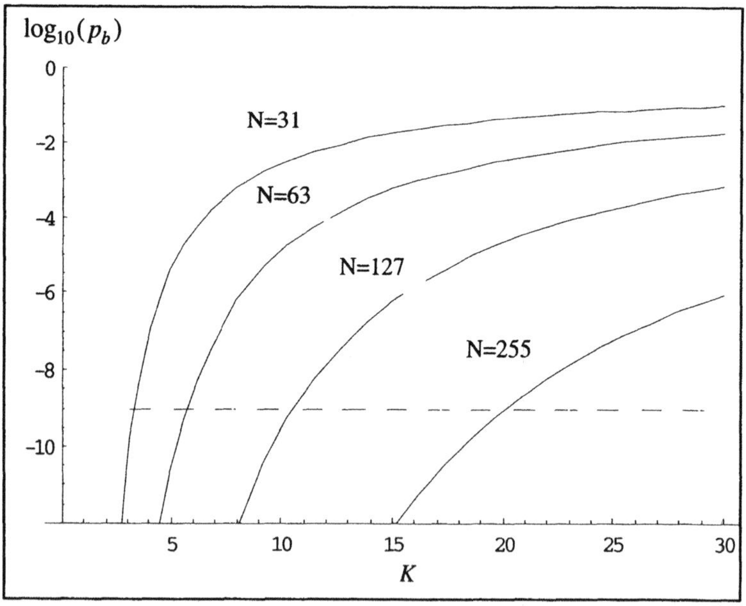

Figure 5. Bit Error rate as a function of the number of users in an asynchronous network.

To compare the proposed system against the more conventional approaches based on unipolar codes, we consider the expected throughput for a given BER target. Let us define the throughput as the maximum allowable aggregate bit rate per channel usage, where the channel usage is the chip slot duration T_c.

In a synchronous network with orthogonal codes of length $N=2^n$ as described in section 3, the optical channel can be shared by 2^n users each with a bit rate $1/2^n T_c$. The aggregate bit rate is thus $1/T_c$ giving the throughput,

$$\eta_{s,b} = 1 \tag{9}$$

In the asynchronous network the system is not interference free, and we have to specify a BER target to define the system throughput. Let be L the argument of the exponential in (8) that guarantees the required BER, then the maximum number of simultaneous users is

$$K_{\max} = 1 + \frac{3N}{2L} \tag{10}$$

and the throughput is

$$\eta_{as,b} = \frac{1}{N} + \frac{3}{2L} \tag{11}$$

To compare (9) and (11) against the conventional use of unipolar codes in optical communications, we consider the results of [3] for asynchronous networks and [5] for the synchronous case. In [3] the authors considered unipolar prime codes, and in [5] a modification of them consisting in an extended set including the original prime codes and the corresponding shifted versions. Prime codes, are binary unipolar sequences of weight P and length $N=P^2$, with P being a prime. The bit error rate in asynchronous networks, obtained with this family of codes, was after some approximations found to be [3]

$$p_b = Q(\frac{P}{\sqrt{1.16(K-1)}}) \tag{12}$$

and for synchronous networks [5]

$$p_b = Q(\frac{P}{\sqrt{K-1}}) \tag{13}$$

where $Q(x) = \frac{1}{\sqrt{2\pi}} \int_x^{\infty} e^{-u^2/2} du$

Let γ bet the argument of Q(.) that guarantees the BER target. From (12) and (13) he maximum number of simultaneous users is

$$\begin{cases} K_{\max} = 1 + \frac{P^2}{1.16\gamma^2} & \text{Asynchronous network} \\ K_{\max} = 1 + \frac{P^2}{\gamma^2} & \text{Synchronous network} \end{cases} \tag{14}$$

and the throughput

$$\begin{aligned} \eta_{as,u} &= \frac{1}{P^2} + \frac{1}{1.16\gamma^2} & \text{Asynchronous network} \\ \eta_{s,u} &= \frac{1}{P^2} + \frac{1}{\gamma^2} & \text{Synchronous network} \end{aligned} \tag{15}$$

Using (9), (11) and (14), we get in the limiting case of long codes, and for BER target $p_b = 10^{-9}$, the following results:

- Unipolar asynchronous $\eta \to 2.3\%$

- Unipolar synchronous: $\eta \to 2.8\%$
- Proposed scheme asynchronous: $\eta \to 7.5\%$
- Proposed scheme asynchronous: $\eta = 1$.

We note that with unipolar codes there are no significant throughput improvements when going from an asynchronous to a synchronous network. This is an expected result, since the motivation of the author in defining the modified prime codes was to increase the number of available code sequences while maintaining the cross-correlation limits. The use of the proposed scheme with bipolar codes provides a dramatic improvement if the network is synchronous. For the asynchronous case, the improvement is far less spectacular but still non-negligible, and in fact with well-designed codes such as Kasami or Gold, we expect a slightly better improvement than the one obtained assuming random sequences.

5. CONCLUSIONS

In this communication we have proposed and analyzed a transmission / reception scheme that allows the combination of non-coherent detection and bipolar codes in optical CDMA. The scheme relies on the optical bipolar coders / decoders proposed in [2], and is basically a direct-sequence spread spectrum transmission system with a non-coherent FSK modulation. The use of non-coherent detection eliminates the need of the complex and sensitive phase control loops, essential in any coherent scheme, and represents thus an important step towards the simplification of receivers. The analysis of the system was done both for synchronous and asynchronous users in the network. The results show that the main feature of bipolar codes arises in synchronous networks, where the use of orthogonal bipolar codes allows a dramatic improvement in the throughput when compared against the conventional sequences. With asynchronous users, there is still some significant improvement although much less pronounced that in the case of synchronous networks.

APPENDIX A

Let us consider the signal defined by (7). We shall without loss of generality consider the detection of user 0, and assume the delay and phase shifts for this signal to be the reference, i.e. $\tau_0 = 0, \phi_0 = 0, \theta_0 = 0$. The signal at the output of the matched filter at T, is (omitting the factor $\alpha\sqrt{P_R}$)

$$\begin{aligned} r_0(T) &= \alpha\sqrt{P_R}\left(a_0^{(0)} NE_p + \sum_{i=1}^{K-1} \left(a_{-1}^{(i)} R_{0i}(\tau_i) + a_0^{(i)} \hat{R}_{0i}(\tau_i) \right) e^{j\phi_i} \right) \\ r_1(T) &= \alpha\sqrt{P_R}\left((1-a_0^{(0)}) NE_p + \sum_{i=1}^{K-1} \left((1-a_{-1}^{(i)}) R_{0i}(\tau_i) + (1-a_0^{(i)}) \hat{R}_{0i}(\tau_i) \right) e^{j\theta_i} \right) \end{aligned} \tag{A.1}$$

where

$$\begin{aligned} R_{0i}(\tau_i) &= E_{p1}^{(i)} \sum_{j=0}^{r} c_j^{(i)} c_{N+j-r-1}^{(i)} + E_{p2}^{(i)} \sum_{j=0}^{r-1} c_j^{(i)} c_{N+j-r}^{(i)} \\ \hat{R}_{0i}(\tau_i) &= E_{p1}^{(i)} \sum_{j=r+1}^{N-1} c_j^{(i)} c_{j-r-1}^{(i)} + E_{p2}^{(i)} \sum_{j=r}^{N-1} c_j^{(i)} c_{j-r}^{(i)} \end{aligned} \tag{A.2}$$

with $\tau_i = rT_c + \varepsilon_i \ (0 \le \varepsilon_i < T_c)$ and

$$\begin{aligned} E_{p1}^{(i)} &= \int_0^{\varepsilon_i} p_\tau(t) p_\tau(t-\varepsilon_i) dt \\ E_{p1}^{(i)} &= \int_{\varepsilon_i}^{T_c} p_\tau(t) p_\tau(t-\varepsilon_i) dt \end{aligned} \tag{A.3}$$

To obtain tractable results we assume that the multi-user interference affecting the components $a_o^{(0)} NE_p$ and $(1-a_o^{(0)})NE_p$ in (A.2) can be modeled as a complex Gaussian process. This is a reasonable assumption for a moderate to high number of simultaneous users. Furthermore we shall use the following assumptions:

- The phase shifts ϕ_i and θ_i are uniformly distributed in the interval $[0,2\pi]$ and are uncorrelated.
- The time delays τ_i, are uniformly distributed in the interval $[0,T]$ and independent.
- The code sequences $\{c_j^{(i)}\}$ are assumed to be independent sequences of i.i.d. random variables taking the values 1 and −1 with equal probability.

The duration of $p_T(t)$ is T_c.

Let us define the real and imaginary parts of the interferences as

$$I_0(T) = \sum_{i=1}^{K-1} \left(a_{-1}^{(i)} R_{0i}(\tau_i) + a_0^{(i)} \hat{R}_{0i}(\tau_i) \right) e^{j\phi_i} = n_{c1} + jn_{s1}$$
$$I_1(T) = \sum_{i=1}^{K-1} \left((1-a_{-1}^{(i)}) R_{0i}(\tau_i) + (1-a_0^{(i)}) \hat{R}_{0i}(\tau_i) \right) e^{j\phi_i} = n_{c2} + jn_{s2} \quad (A.4)$$

It is easy to verify that with the assumptions pointed out we have

$$(n_{ci}, n_{si}) \propto N(0,(K-1)T_c^2) \quad i = 0,1 \quad (A.5)$$

Because of the symmetry of the problem, we may assume without loss of generality $a_0^{(0)} = 1$, and compute the probability of bit error for this case. We get an error decision if

$$\left| NT_c + n_{c0} + jn_{s0} \right|^2 < \left| n_{c1} + jn_{s1} \right|^2 \quad (A.6)$$

and then using the well-known results on the probability of a Rice random variable exceeding another one [14] we obtain

$$p_b = \frac{1}{2} e^{-\frac{3N}{2(K-1)}} \quad (A.7)$$

REFERENCES

[1] Special Issue on Multiwavelength Optical Technology and Networks, IEEE J. Light.., vol. 14, June 1996.

[2]K. Kitayama, "Code division multiplexing lightwave networks based upon optical code conversion", ", IEEE J. Select. Areas Commun., vol. 14, pp. 1309-1319, Sept. 1998.

[3] P. R. Prucnal, M. A. Santoro and T. R. Fan, "Spread-spectrum fiber optic local area network using optical processing", IEEE Journal of Light. Tech., pp.547-554, May 1986.

[4] J. A. Salehi and C. A. Brackett, "Code division multiple access techniques in optical fiber networks – Part II: System performance analysis", IEEE Trans. Commun, pp.834-842, Aug. 1989.

[5] W. C. Kwong, P. A. Perrier and P. R. Prucnal, "Performance comparison of asynchronous code-division multiple-access techniques for fiber-optic local area networks", IEEE Trans. Commun, pp.1625-1634, Nov. 1991.

[7] J. K. Tang and K. Ben Letaief, "Optical CDMA communication systems with multi-user and blind detection", IEEE Trans. Commun., vol. 47, pp. 1211-1217, Aug. 1999.

[8] M. E. Marhic, "Coherent optical networks", Journal of Light. Tech., pp. 854-864, May/June 1993.

[9] N. Wada and K. Kitayama "A 10Gb/s optical code division multiplexing using 8-chip optical bipolar code and coherent detection", Journal of Light. Tech., pp. 1758-1769, Oct. 1999.

[10] W. Huang and I. Andonovic, "Coherent optical pulse CDMA systems based on coherent correlation detection", IEEE Trans. Com, pp.261-271, Feb. 1999.

[11] F. Chung, J. Salehi and V. Wei, "Optical orthogonal codes: design, analysis and applications", IEEE Trans, Inf. Theory, pp. 595-604, May 1989.

[12] A. S. Holmes and R. Syms, "All-optical codes using "quasi-prime" codes", IEEE Journal of Light. Tech., pp.279-286, Feb 1992.

[13] M. Chang and M. E. Maric, "Fiber-optic ladder networks for inverse decoding coherent CDMA", Journal of Light. Tech., pp. 1952-1962, 1992.

[14] J. G. Proakis, Digital Communications, MacGraw Hill 1989.

Access Control in Shared Access Networks Supporting Internet DiffServ

J. D. Angelopoulos, N. Leligou, Th. Orphanoudakis, G. Pikrammenos,
J. Sifnaios and I. S. Venieris

*National Technical University of Athens,
Electrical and Computer Engineering Department,
GR15753, Athens, Greece, e-mail: jangel@cs.ntua.gr*

Key words: MAC, HFC, PON, DiffServ, QoS.

Abstract: Shared access networks such as hybrid fiber/coaxial and passive optical networks have emerged as promising ways to reduce the cost of the transition to a broadband access infrastructure and provide a graceful upgrade path towards the photonization of the local loop. The TDMA multiplexing of traffic entering such a system is governed by the MAC (Medium Access Control) protocol, which arbitrates the allocation of bandwidth to the shared feeder. At the same time the need to integrate telecom services presenting demanding quality requirements with plain Internet best-effort services over the same infrastructure, brings new issues to the design of such an access mechanism. The MAC protocol as the only arbiter of the upstream bandwidth directly affects the Quality of Service (QoS) provided to each upstream traffic flow and must meet several constraints. Such constraints include the adequate speed of operation exploiting in the highest degree the speed of H/W implementation, flexibility to support efficiently the largest number of services and applications offering an adequate number of QoS classes like the differentiated services (DiffServ)and independence of higher layers, protocols and future extensions to traffic management specifications. The implementation of a mechanism targeting these goals and its evaluation using computer simulation are presented in this paper.

1. INTRODUCTION

The access part of telecommunication networks also known as the "local loop" is characterised by low traffic intensities and high cost per traffic volume. It is typically amortized by very few customers and the last drop by just one. Thus it is very sensitive to cost and experiences much longer investment life-cycles. To reduce the cost while retaining the evolutionary prospects, tree-shaped topologies exploiting a shared feeder fiber are a promising solution. Typical examples are bi-directional Hybrid Fiber/Coaxial (HFC) networks (an evolution of CATV systems) as well as passive optical networks (PONs). Both architectures rely on shared optical fibers in the feeder section while the existing copper infrastructure remains on the home side at least for the introductory phase when fiber will not go into the last drops.

In both systems, sharing of the upstream channel (from customer to network) is usually effected through TDMA multiplexing. A Medium Access Control (MAC) protocol is employed to arbitrate the access to the slotted common medium by allocating slots to customer terminations according to demand [5], [6].

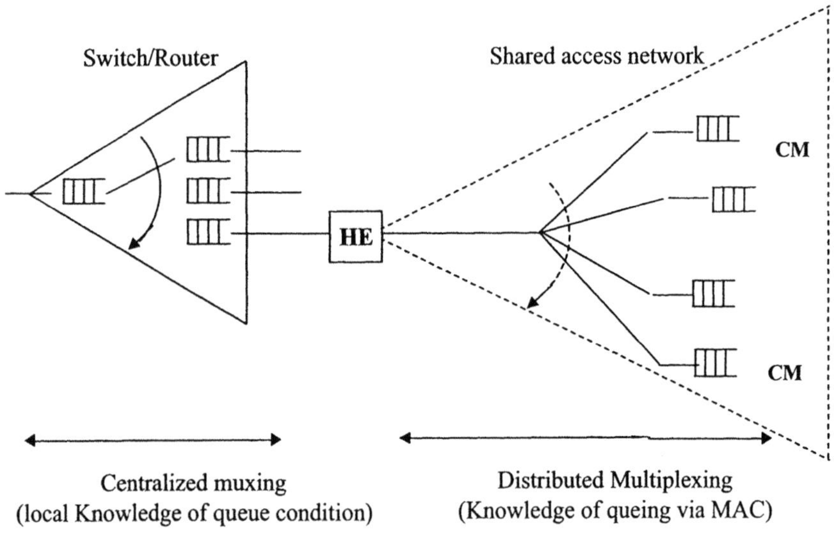

Figure 1. Conceptual view of distributed access multiplexing

This results in a distributed queuing system characterised by the long time required to pass control information from the queuing points to the service controller residing at the head-end. The allocation of upstream slots in a tree-topology access system is based on a reservation method which allows to dynamically adapt the bandwidth distribution to traffic fluctuations. The MAC controller collects access requests and then allocates the upstream slots by sending access permits. It is important to note that the service policy of the MAC governs the distributed multiplexing from a central point situated miles away at the head-end. This has the important implication that regarding the delay control, the acquisition of arrival information and buffer fill levels includes a considerable delay element. This delay element does not exist in centralized multiplexing schemes, operating typically at a switch queuing point. In contrast, drop policies must be distributed over all network terminations (cable modems) where the flow identity and fill levels are known. Because of the much larger reservation delay and statistical behavior of the aggregations from many customers, special care must be taken to safeguard QoS to sensitive traffic. This requires a prioritization scheme integrated in the MAC function of PONs and HFC systems as was already identified in [5]. This must be in line with the recent advances in networking that include the two major approaches in offering multi-service broadband access enriched with differentiated QoS support namely the categorization of traffic classes by the ATM Forum [8] and the differentiated services (DiffServ) strategy of IETF [1]. In the following sections the implementation and performance of an HFC MAC aligned to the DiffServ architecture will be presented. Similar design considerations apply to PON MAC protocols with some adaptation to the higher line rates of a PON.

2. DESIGN OF A MAC FUNCTION FOR A SERVICE INTEGRATED MULTILAYER HFC ACCESS NETWORK

From a traffic engineering point of view, the tree-shaped access system exhibits a very different behaviour between the upstream (from customers to core) and the downstream (from core to the customers) direction. In the downstream, the broadcast nature of the medium creates replicas of the signal (in the passive splitters of PONs or the taps of HFC systems) giving rise to privacy and security issues. These are typically dealt with by means of encryption, which is out of the scope of this work. In the upstream on the other hand, the MAC control function effects multiplexing / concentration of the traffic from all active modems as typically occurs in the output ports of any switching node (*Figure 1*). However in this case it is characterized by the distributed nature of the queuing points and the additional difficulties in the exchange of control information. Failure to recognize the packet precedence and behave accordingly would result in loss of QoS for any sensitive flow under high load. To apply any scheduling or priority discipline requires the correlation of the traffic from all multiplexed sources going to the one common egress point of the system.

Traffic categorization may be connection based following the ATM approach or flow based following the DiffServ approach of the IETF. The system under consideration employs a TDMA slotting designed for ATM cells. Such an ATM slotting complies with the approach followed by the main standards bodies, like DAVIC 1.3 and 1.4, DVB/ETSI ETS 300 800, (now incorporated in DAVIC), and IEEE 802.14 [4]. Regarding PONs the ATM slot size is used in G.983 [9]. So we will consider the ATM size slots as a quantum of the MAC assigned bandwidth allocation. To accommodate IP frames, several slots are successively assigned by the MAC. However, even if IP is not offered over ATM, the legacy of short slots can be used to advantage in the context of the DiffServ architecture over a slow shared link such as HFC. Namely, it allows to suspend the transmission of a lower priority packet, on the boundary of a slot (cell), and transmit delay-sensitive packets before resuming the transmission of the low priority one. In addition, fixed slots are easier to handle in a H/W based MAC.

Logically the MAC protocol functions similarly to a centralized multiplexer i.e. connections or IP flows with demanding QoS must be identified and receive properly differentiated treatment from the plain best effort traffic. This is accomplished in the cable modem or the ONU during packet classification by placing the corresponding cells in the high priority queue which will subsequently place higher priority requests and activate the higher priority permit allocation MAC algorithms.

To reduce complexity in this cost-sensitive residential access system, services are grouped into behavior aggregates (classes) with a similar set of requirements providing scalability and flexibility. Given the fluidity of service class definition and the fact that the MAC is implemented mainly in H/W, it was deemed adequate to implement just 4 priority classes in the MAC, leaving finer aggregations and more elaborate forwarding policies, to be implemented at the egress of the HFC system. All that is required from the MAC is not to deny quality to any groups of flows. The basis of this approach is the use of access priorities in the reservation system which can be programmed to fit with required traffic descriptors by means of S/W programming of the mapping of flows to priorities.

The characteristics of the four aggregation levels/priorities are the following:

- The high priority is devoted to delay-sensitive periodic CBR traffic.This class targets services with very strict delay requirements, which undergo strict traffic profile control (traffic conditioning) such as CBR ATM connections and the EF (Expedited Forwarding) service [2].

- The second priority level is devoted to real-time variable rate flows, such as video services or VoIP and it is provided with a minimum guarantied rate by means of pre-allocated permits and peak rate policing for guaranteed QoS. MAC exercises a policing function by rate checking before issuing the permits. This is based on credit allocation at the time of subscription or connection set-up. In the DiffServ context it could be used for the top AF (Assured Forwarding) class.
- The third priority is devoted to data services with higher requirements than best-effort. The traffic profile control assumed for this class aims at minimizing the loss of packets and the disturbance to other traffic. The credit scheme is used to guarantee a minimum rate (while credits last) while traffic exceeding this limit is relegated to the 4th priority permit generation. The 3rd priority mechanism is suited to the support of all four or the lower three AF classes [3] as well as ABR connections in the ATM context that also need separate queuing and information about buffer levels in order to possibly integrate the closed loop control into the MAC function as described in. (Drop policies can be independently applied at the modem queuing points).
- The fourth priority is reserved for plain best-effort services such the UBR as well as traffic which employ loss based flow control at the TCP level and can be very disruptive to the other classes when sharing the same queue.

The last three priorities employ reservation while the first unsolicited permits. The reservations operate by means of three request fields totaling 8 bits. Two bits are used for the 2nd class and three bits for the other two creating a byte which is piggy-backed in every upstream slot. This byte uses the place of the HEC field of the ATM cell which is not needed for cell delineation since an additional synchronization preamble is employed because of the burst mode operation. The total length of the upstream slot has a size of 64 bytes accommodating except of the cell payload a synchronization preamble. The frame structure in the downstream direction is dictated by several mandates of the physical layer (synchronization, flexibility over many modulation constellations and rates adapting to plant conditions, interleaving, Reed-Solomon FEC etc). These issues are covered in the relevant standards, e.g. [4], and will not be discussed here since we focus on the traffic multiplexing issues. For the operation of the MAC all that is important is that a number of permits is provided in a periodic basis corresponding to the number of upstream slots over the same period.

The reservations in the HFC MAC are usually assisted by contention on special reservation mini-slots [4] in addition to the piggy-backed requests. This is needed for the announcement of the first arrival of a burst since the piggy-back mechanism is not self-starting but relies on the existence of previous traffic for the announcement of new arrivals [6]. However the MAC departs from this approach in two points: first it provides for all three simultaneous requests (one from each queue) as explained above, irrespective of which queue provided the cell, and second it employs round robin polling instead of contention. Simple polling is considered adequate since the system targets an advanced stage of deployment when the HFC take-up will be significant and the customer clusters will be smaller with much higher bandwidth needs. In PONs, polling is preferred because there is a high number of slots contained in the round trip making collision resolution impractical.

The multiple requests are the tool for the higher QoS capabilities of the system. They are necessary if higher priority traffic is to be quickly made known to the head-end. This feature enables the algorithm at the head-end to offer precedence in the permit allocation to the high priority on a global basis and not just among the cells of the same termination, which is of a very limited value.

3. IMPLEMENTATION OF THE MAC CONTROLLER

The MAC controller was implemented on the Access Network Adaptation board. Since the MAC considered a sub-layer of the physical layer and is closely related to the physical layer functions (TDMA, framing, etc.) the permit generation function was selected to be integrated with the downstream framer component, which was responsible for the construction of the downstream frame including the transmission permits. The whole design was programmed and placed on a Field Programmable Gate Array (FPGA) chip. This decision made possible the re-design of the implementation to adjust to modified frame structures, rates etc. An embedded processor (On Board Controller) was used to calculate and modify all non-real time parameters such as the programmed (pre-allocated or provisioned) bandwidth distribution, which may vary with time due to the switch on or off of the modems. An external static RAM chip was used for buffering necessary protocol information in addition to the available on-chip memory provided by the FPGA, which was exploited to keep state information and speed-up the permit generation process. The MAC algorithm works as follows:

A list of 512 permits is employed, which has been prepared by the OBC on the basis of subscription data for the scheduling of all pre-arbitrated permits. The OBC executes the Call Admission Control (CAC) scheme and writes periodically permits for the 1st and 2nd priority connections at the peak and the minimum guarantied rate accordingly. Permits are also inserted for polling requests by modems that have not established a 1st or 2nd priority connection. A 3-bit field is foreseen in the permit format in order to indicate the queue (and the corresponding QoS class) targeted, facilitating internal calculations. This list is cyclically read out by the MAC controller H/W, and the permits sent downstream. It worths noting that the format of the permits sent downstream differs from that written in the list in the sense that they include no coloring but only the modem ID and the modem is responsible for selecting the appropriate queue for transmission in an increasing order of priority. Techniques to space the permits in the list are given in [7]. This list is stored on the external SRAM chip. At the end of each cycle, the embedded controller updates also a list of credits for the 2nd and 3rd priority connections. These credits are used for policing 2nd priority connections at the contracted peak rate and guaranteeing the minimum calculated bandwidth of 3rd priority connections as described above. Since ATM signaling is also supported, the permit and credit lists can be updated dynamically to add new connections using a second copy.

The three lower priorities are serviced dynamically on the basis of requests by filling in locations whenever they are left empty in the permit list (representing unallocated bandwidth). The requests per queue/priority arriving at the MAC controller in the head-end, are used to update the 2 outstanding request totals for each modem and queue, by incrementing the relevant counters by the amount of new cell arrivals. At the same time the downstream permit positions in the frame are filled with permits produced by an engine scanning in a round robin fashion the outstanding request counters and reducing them by one for each permit scheduled. The higher priority counters are inspected first and only if all are empty the same process is repeated for the immediately lower priority. To expedite the process, flags are used to quickly detect and skip all empty locations. The MAC H/W subtracts the credits as it issues permits and stops serving any modem queue that exceeded its allocated apportionment. This policing action guarantees that malicious users can not disturb complying traffic. The credit check is used differently in the 3rd priority. When credits are exhausted, left-over requests are added to the 4th priority ones thus a minimum rate is guaranteed but any excess is considered plain best-effort, in accordance with AF rules. The operation of the permit scheduler and the execution of the MAC algorithm is depicted in *Figure 2*.

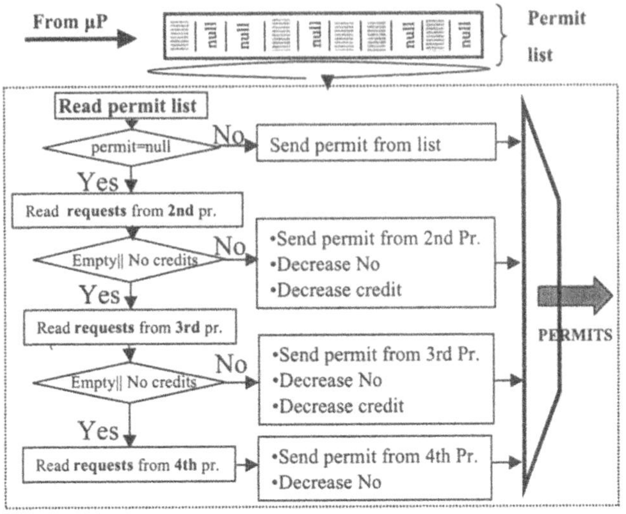

Figure 2. Operation of the permit scheduler.

4. PERFORMANCE ASSESSMENT

For the evaluation of the above MAC mechanisms a computer simulation model was created using the PTOLEMY platform for modeling the MAC function under certain loading conditions. The scenarios used 10 cable modems loaded with uniform traffic per priority for each modem. Each source used a common ON-OFF model, generating traffic at the slot/cell level, with on and off periods exponentially distributed. The upstream channel rate was set as peak rate. There was no limit set to credits, since the policing function would only be meaningful to demonstrate with malicious sources that was out of the scope of these tests, since the results are determinist. The duration of the runs was 1.5 million slots and the upstream slot duration was used as a time unit for the simulations (170.6 µs).

Figure 3 depicts the probability distribution function (pdf) of the access delay for the three lower priorities (2nd, 3rd, and 4th) under a total load of 85%. The 1st priority list was using 10% of the load for CBR traffic (or virtual leased lines by EF in the DiffServ model of usage), with the other 75% equally distributed among the other 3 priorities. Only the variable part of the delay is shown, as the fixed round trip time (about 4 slots) required for the modem request (upstream) and the permit response (downstream) is not counted.

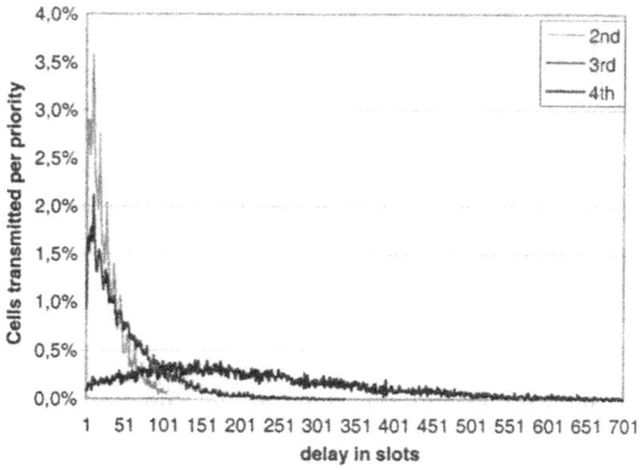

Figure 3. Pdf of access delay (total load 110%).

The 1st priority is not presented since it exhibits deterministic behaviour, as the transmission permits are periodically pre-programmed in the list [7], and thus, the delay never exceeds the fixed permit distance. The delay advantage provided by the prioritization leads to almost all the slots of the 2nd priority accessing in less than 250 slots (i.e. 43ms) while those of the 3rd in less than 350 slots (60ms). Of course there is no bound for the 4th priority which can exceed any limit depending on the total loading.

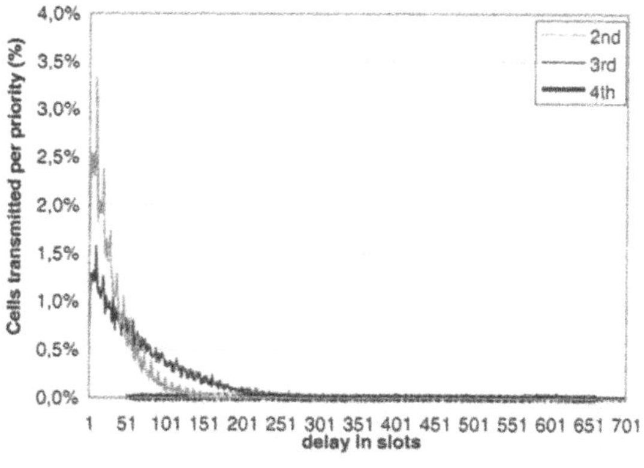

Figure 4. Pdf of access delay (total load 85%).

The pdf of delay under a higher offered load of 110% is shown in *Figure 4*. The list contained 10% unsolicited CBR traffic (1st priority), as the 2nd and 3rd priority sources offered a load equal to 30% of capacity, and 40% the 4th. The first two priorities do not exhibit any significant behaviour difference with the previous scenario, since they encounter a lightly loaded medium as before. The effect of prioritization is exactly to hide the presence of any lower priority traffic, which is prevented from competing against the protected sensitive traffic. The effect on the 4th priority is very strong since it is throttled down to the 30% of capacity left over from the other three. The occurred delay distribution is thinly spread till theoretically infinite delay. This can not be shown in the simulation results since buffers overflow no matter how much memory is allocated. The average delay is not bounded for the 4th priority, as shown in *Figure 5*.

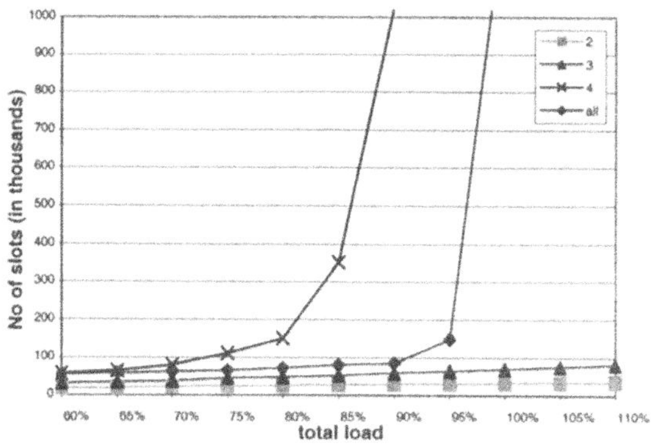

Figure 5. Mean delay vs. total load.

Again 10% of the upstream bandwidth was pre-allocated by means of programmed permits in the permit list as necessary for the polling of new bursts and the rest equally distributed among the priorities and modems for several values of the total offered load as depicted in *Figure 5*. The delay for the 2nd and 3rd priorities increases very smoothly with the load leaving the 4th priority to suffer the congestion. The 4th priority delay increases asymptotically towards the 100% line, as is typical of any queuing system.

At the same figure we can observe the behavior of the system when only two priorities are valid: 1st and "all". The first priority ramains as it was, while the three others are merged to one, the "all". The load balance remains 10% for the 1st while the "all" priority occupies the missing part of the total load represented at the horizontal axis. It is obvious that the "all" priority behaves better than the 4th priority and worst than the 2nd and 3rd. It behaves better than the 4th priority because there are no more priorities that could block the transmission in favor of a higher priorities. At the other hand, it behaves worst than the 2nd and 3rd priorities because it has 3 times their stand alone traffic profile: when 2nd and 3rd priority occupy 20% of the total load each, the "all" priority occupies 60%. The delay is increased grammically with the total load until the 90% point, from where the systems starts to saturate. The "all" line seems to increase asymptotically to the 100% limit.

However, the sources used do not model higher layers where the typical congestion avoidance of TCP would exercise flow control in response to losses and reduce the rate, while re-transmitting lost packets. So the system thanks to prioritization can guarantee the performance enjoyed by sensitive traffic while exploiting any unreserved bandwidth for the support of best-effort traffic.

5. CONCLUSION

Tree-shaped shared medium access networks such as PONs and HFC effect a distributed multiplexing function which concentrates traffic from many users and many services with diverse requirements. To be able to guarantee that the QoS of sensitive traffic will not be disturbed by best-effort data traffic requires embedding differentiated support for flow aggregates with common requirements. The architecture considered in this paper is quite advantageous in this context and a MAC protocol taking advantage of this approach was implemented and evaluated with satisfactory results.

6. ACKNOWLEDGEMENT:

The work presented in this paper was partially funded by the EU ACTS project AC327 "AROMA". The opinions appearing here are those of the authors and not necessarily of the other members of the AROMA consortium.

7. REFERENCES

[1] IETF, Differentiated Services Working Group, RFC 2475 "Architecture for Differentiated Services", December 1998

[2] Van Jacobson, Kathleen Nichols, Kedarnath Poduri, Internet Draft, draft-ietf-diffserv-phb-ef-02.txt, "An Expedited Forwarding PHB", February, 1999

[3] Juha Heinanen, Fred Baker, John Wroclawski, Internet Draft, draft-ietf-diffserv-af-06.txt "Assured Forwarding PHB Group", February, 1999

[4] IEEE 802.14 MAC Draft(R3) Specifications, May 1997 (not publicly available).

[5] J.D.Angelopoulos, G.C. Boukis, I.S.Venieris, "Delay priorities enhance utilization of ATM PON Access Systems", *Computer Communications Journal*, Elsevier, Vol. 20, No. , December 1997, pp. 937-949.

[6] J. D. Angelopoulos, Th. Orphanoudakis, "An ATM-friendly MAC for traffic concentration in HFC systems", *Computer Communications Journal*, Elsevier, Vol. 21, No. 6, 25 May 1998, pp. 516-529.

[7] J. D. Angelopoulos, N. I. Lepidas, E. K. Fragoulopoulos, I.S. Venieris,"TDMA multiplexing of ATM cells in a residential access SuperPON", *IEEE Journal on Selected Areas in Comm.*, Special issue on high capacity optical transport networks, Vol. 16, No. 7, September, 1998.

[8] ATM Forum Technical Committe, Traffic Management Group, "ATM Forum Traffic Management Specification Version 4.0" Doc. 95-013R11, March 1996.

[9] ITU-TS Rec. G.983, "High speed optical access systems based on PON techniques" SG-15, Feb. 1998.

Network Controller Procedures in SONATA, A Large Scale All-optical WDM Network

A. Bianco, E. Leonardi, M. Mellia, F. Neri
Dipartimento di Elettronica, Politecnico di Torino, Italy
{bianco,leonardi,mellia,neri}@polito.it

Abstract This paper describes the network architecture of an optical WDM network named SONATA, which has been proposed and will be demonstrated in the context of the European Union ACTS Program. In this nation-wide all-optical network, end-terminals access a single passive routing node via PONs using a TDMA/WDMA access scheme based on slot reservations. The centralized network controller runs resource allocation algorithms in order to avoid conflicts among end-terminals. Since the complexity of optimal solutions to the resource allocation problem at the network controller is very large, we propose in this paper simple heuristic algorithms to solve the network control problem, and study their performance through simulation. Results of the design of the signaling protocol for SONATA are also reported.

1. INTRODUCTION

The project SONATA aims at avoiding the need for large and fast switching electronic nodes in a high-speed nation-wide network. To reach this goal, the network structure and the layer architecture within the network have been drastically simplified. The new proposal consists of a single-layer network platform for end-to-end optical connections, able to serve a very large number of terminals for both business and residential customers over a nation-wide geographical area.

Previous proposals of WDM optical networks can be roughly classified into two broad categories: *single-hop systems*, providing a fixed connectivity (broadcasting is often considered, as in the case of networks built

*This work was partially funded by the European Union ACTS Program within Project AC351 SONATA.

around a passive star coupler,) among terminals, which run access protocols to share transmission resources (see, e.g., [1]), and *wavelength routing systems*, usually based onto general mesh topologies, in which optical cross-connects can be programmed to setup optical data pipes between terminal pairs (see, e.g., [2] and [3]). SONATA attempts to combine the best features of these two approaches: on the one hand, it keeps the architectural simplicity of single-hop systems, and terminals actively contribute to share the transmission resources. On the other hand, the possibility of dynamically modifying the end-to-end connectivity provided by the network is introduced to efficiently use network resources. Note also that the latter feature is one of the major strenghts of optical networks with respect to electronic networks.

In the SONATA network, a large number of end-terminals, which, depending on the network usage, can be street cabinets, IP routers, LAN switches, or workstations, are grouped in a Passive Optical Network (PON) infrastructure. The network topology is based on a centralized Passive Wavelength-Routing Node (PWRN) with N input and N output ports, which provides full connectivity among PONs via a single dedicated wavelength channel for each pair of PONs. Wavelength converters arrays are used at the PWRN in order to dynamically increase the channel capacity among pairs of PONs, as described below.

This multiple-access network is based on WDMA/TDMA protocols and exploits time and wavelength agility at terminals in a very simple network structure, where the single PWRN node provides passive routing functions and actively controlled wavelength conversion. The network control is centralized, and its primary goal is to assign time and wavelength resources to terminals: a signaling protocol that allows terminals to require both connection-oriented and connectionless services has been designed. The resource assignment performed at the Network Controller (NC) avoids any conflict among transmitters and receivers when transmitting data.

In SONATA, the physical switching function is removed from network nodes and distributed at terminals. Although a centralized network control is required, and active wavelength conversion is performed inside the network, we refer to the SONATA network structure as a "switch-less" network, in the sense that neither purely electronic switching nodes or cross-connects (telephony, IP, ATM, SDH), nor optical cross-connects (except for the wavelength routing node) are required within the network. Moreover, the network is completely buffer-less. This approach provides major network architecture simplifications and hardware reductions.

We do not tackle issues related to the SONATA network feasibility in this paper: we do not discuss neither the components that should be used nor the physical limitations that should be taken into account when dimensioning the network. All these issues have been deeply analyzed in the SONATA project, and are discussed in [4, 5, 6].

The paper focus is on algorithms that must be executed at the NC to solve the resource allocation problem. We first provide a more detailed description of the switch-less network architecture. Then, we discuss the time/wavelength resource allocation problem at the Network Controller. It can be shown that an Integer Linear Programming (ILP) formulation of the problem is possible, and that its optimal solution is in general NP-hard. We therefore propose a simple heuristic to solve the problem of resource allocation: we decouple the time dimension from the wavelength dimension, thus splitting the problem into two sub-problems: the scheduling of terminal requests in the time domain given a PON-to-PON channel assignment, and the design of the logical topology, i.e., the assignment of wavelengths to PONs via a proper setting of wavelength converters. The performance of the proposed algorithms is studied via simulation.

2. NETWORK ARCHITECTURE

A more detailed description of the network architecture is provided in this section. Recall that the SONATA switch-less network has the target of performing the concentration/distribution, switching and routing functions within a single network layer by providing end-to-end optical connections between a large number of terminals over a wide geographical area.

The structure of the switch-less network is depicted in Fig. 1. Terminals equipped with fast tunable transmitters and receivers are attached to Passive Optical Network (PON) infrastructures. Each PON is directly connected to an input and an output port of the single Passive Wavelength-Routing Node (PWRN). The PWRN has N input and N output ports, i.e., it is an $N \times N$ wavelength multiplexer. Logically, the behavior of the PWRN is such that wavelength λ_k at input i is routed to output $j = |i+k|_N$. Hence wavelength channels $0, 1, 2, \ldots, N-1$ on input port i lead to output ports $i, i+1, i+2, \ldots, i-1$, and, on output fiber j, wavelength channels $0, 1, 2, \ldots, N-1$, carry information originated at input $j, j-1, j-2, \ldots, j+1$.

Each wavelength is shared between all terminals belonging to the same PON. A TDMA control protocol is therefore required. A global synchronization is available, and transmission is organized in WDMA/TDMA

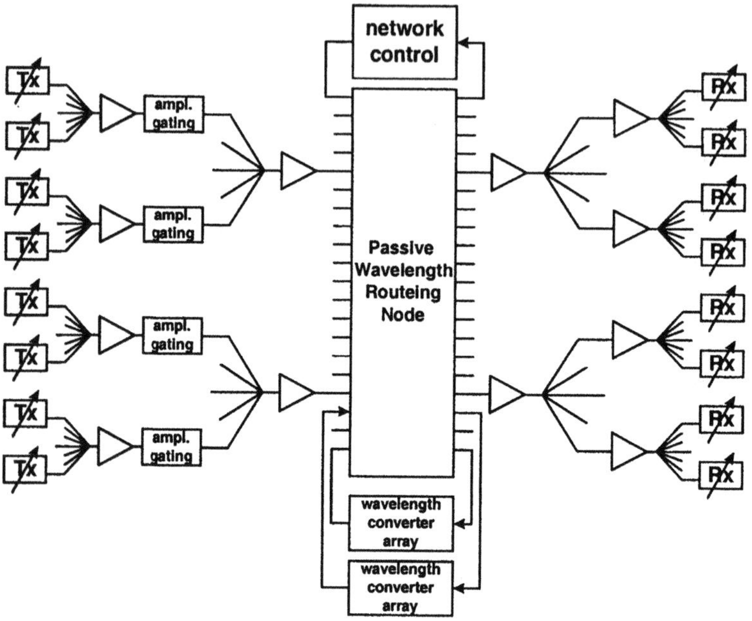

Figure 1 SONATA network architecture

frames. Each frame is composed of F fixed-length slots. Any terminal wishing to communicate with another terminal simply tunes, in the allocated time slots, its transmitter and receiver to a known wavelength carrying multiplexed traffic through the PWRN between the pair of PONs to which the transmitting and receiving terminals are attached. Only one tunable transmitter/receiver pair per terminal (plus a fixed transmitter/receiver for signaling) therefore permits data communication between a given terminal and all other terminals in the network, regardless of the network size. Note that multicast transmissions to several terminals belonging to the same PON can be supported.

Although all connections can be made directly through the PWRN, as the traffic volumes grow towards the capacity limit of the network, a high degree of flexibility in allocating capacity can be obtained with actively controlled wavelength converter arrays. As shown in Fig. 1, these devices can be added, as required, around the PWRN, and connected to a certain number (N_d) of its auxiliary ports (called "dummy ports"). On each PON, a wavelength channel is available to reach a particular dummy port, hence a wavelength converter array. From each array, one wavelength channel is provided to reach every output PON. Hence information from PON i to PON j can be routed on the direct channel

$|j - i|_N$, and on up to N_d additional channels going through wavelength conversion. In this way a variable number of wavelengths, depending on traffic requirements, can be dynamically allocated between a pair of PONs. Thus, the network structure depicted in Fig. 1 provides a full mesh topology between PONs through direct connections, and a large number of additional connections through wavelength converter arrays.

Additional N_c ports of the PWRN interconnect network terminals with the centralized Network Control (NC) device, which is responsible for allocating time slots and wavelength channels to terminal requests.

The optimal fisical solutions for a "switch-less" network depend on the size of the network itself. Triangles in Fig. 1 represent optical wide-band amplifiers. Amplifier gating blocks are used to eliminate (using SOA gates) noise from wavelength channels where no transmission occurs (see [4]). Global synchronization is obtained with a distribution of reference lines, and with the implementation of ranging functions to compensate for variable propagation delays between terminals and the PWRN.

As discussed in [4], a feasible set of characteristics for a large scale switch-less network, which can provide a communication infrastructure to terminals at the customer premises distributed nationwide, could be the following (assuming $N_c = 1$):

- $N_t = 20$ million terminals
- 1,000 km maximum distance between terminals
- 400 PONs with 50,000 terminals each
- up to $N_d = 400$ wavelength converter arrays with 400 wavelength converter modules each
- 801 × 801 wavelength router with W=801 wavelength channels on each fiber
- 10 Mbit/s total average rate per terminal
- 622 Mbit/s maximum rate per terminal
- 200 Tbit/s maximum network throughput
- $F = 1,000$ slots per frame
- $10\mu s$ of slot time, with $1\mu s$ devoted to guard time

The drastic simplification of the network structure eliminates critical complex network components (note that the infrastructure is mainly passive), thereby increasing network reliability. No failure protection is

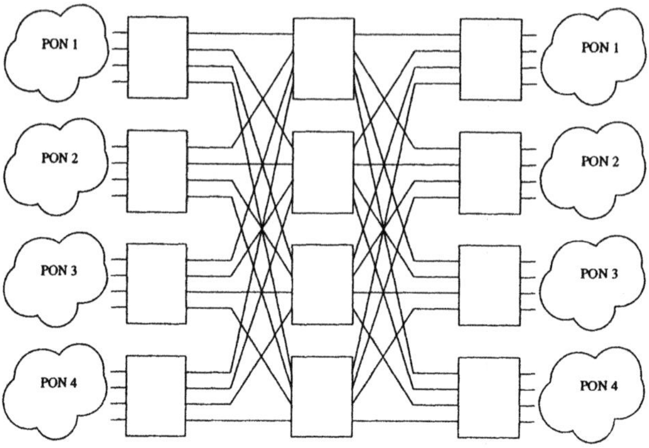

Figure 2 Three stage equivalent of the programmable portion of the SONATA network

however provided in the architecture of Fig. 1, and critical components (for example the Network Controller) may have to be duplicated for reliability in real implementations.

Logical Topology. The first N_p ports of the PWRN provide a "wired" (fixed) full-mesh connectivity between PONs, while the additional N_d dummy ports add a "programmable" connectivity between PONs. Fig. 2 shows the equivalent logical topology provided in the programmable portion of the PWRN in the case $N_d = N_p = 4$. Terminals in PON i can reach any of the N_d wavelength converter arrays by tuning to the proper wavelength channel. This tuning capability corresponds to the first λ-switching stage in the logical topology. The individual wavelength converters can be configured in order to route input information to any output PON, leading to the second λ-switching stage in Fig. 2. The receivers in each PON can tune to any of the N_d wavelengths routed to the PON by the wavelength converter arrays, leading to the third λ-switching stage. The resulting logical topology is a three-stage $\lambda - \lambda - \lambda$ switch, i.e., a Clos interconnection network. It is well known from the Slepian-Duguid theorem [7] that this three-stage interconnection structure is rearrangeable non-blocking when $N_d \geq N_p$, as it is normally the case for SONATA. Note that the interconnection between PONs is non-blocking, while the interconnection between terminals is blocking, due to the multiplexing of many terminals in each PON (at most 800 out of 50,000 terminals can receive/transmit in each slot) of a full-size network.

If we keep considering only the programmable portion of the network, the possibility of allocating transmission requests in different slots within the time frame can be viewed as a further switching stage in the time domain. Overall, the programmable portion of the switch-less network can be logically described as a five-stage $T-\lambda-\lambda-\lambda-T$ switch, which can be re-configured in every time slot.

Since time is divided into time slots, and each time slot can be used in a particular wavelength channel by only one of the terminals connected to the same PWRN port, a significant complexity is required in the NC operations. Considering a SONATA network deployed to its full capacity, the NC has to keep track of the state of $(Nc + Nd)^2 \times F = 640,000,000$ time slots in every frame. Further constraints on the controller are due to the fact that the single transmitter/receiver pair has to be shared by all the connections which the terminal keeps active at the same time; more precisely, to permit a transmission from terminal s to terminal d in slot k, we must ensure that s is not already transmitting in k, and also that d is not already engaged in a reception in k. This requires in principle $2 \times N_t \times 1000 = 40,000,000,000$ state variables for each user.

3. RESOURCE ALLOCATION ALGORITHMS IN SONATA

The main task of the Network Controller is to allocate resources (time slots) to end-terminals, choosing a proper wavelength among the available ones.

Although globally optimal algorithms to allocate resources are possible, we propose in this paper a much simpler approach to resource allocation at the Network Controller, as the complexity of optimal control algorithms is so large that their implementation is impossible, given the very large network size. Moreover, the amount of signaling data that has to be sent to each terminal if optimal algorithms are used is very large, as mentioned in [5].

Our simple approach is based on the decoupling of the time dimension from the wavelength dimension. This means that we split the resource allocation problem into two sub-problems: *scheduling* of terminal requests in the time domain given a PON-to-PON channel assignment, and *logical topology design* of the network connectivity by properly assigning wavelengths to PONs via the wavelength converter arrays.

While scheduling is a task which must be done slot-by-slot, the logical topology design is assumed to be performed at a lower rate. Aiming at a sub-optimal approach with a limited complexity, we assume that the wavelength converters are reconfigured only once in a while (e.g., at frame boundaries) and that the scheduling algorithm operates on a given fixed logical topology.

We analyze in the sequel the two-steps of the resource allocation algorithm:

- Scheduling: first-fit allocation of slots given a fixed network connectivity;
- Logical Topology Design: reconfiguration of the network connectivity via wavelength converter arrays.

Scheduling. The scheduling algorithm is executed at the Network Controller on the basis of terminal requests received via signaling messages. Terminals can request slots according to two different modes:

- "persistent" request: an integer number of slots in each frame is assigned at connection setup; this number can be changed (either increased or decreased) during the connection duration
- "non-persistent" request: a (possibly large) amount of slots is assigned to a datagram communication, without time constraints in terms of number of involved frames

Obviously, the two modes refer mainly to CBR (or slowly varying VBR) connection-oriented services, and to datagram services, respectively.

When terminal s belonging to PON S sends a request to the NC in order to communicate with terminal d belonging to PON D, the sequence of steps executed at the NC is the following.

- If terminal s requires the set up of a new connection, the NC looks for a candidate wavelength channel by scanning a linked list of wavelength channels assigned to the pair of PON $S \to D$. The first element of the linked list is always the "wired" channel among PONs $S \to D$, and the remainder of the list is ordered in increasing order of channel utilization. The new allocation is attempted on the first channel of the list that satisfies the condition that the number of free slots exceeds by a given percentage b the number of requested slots.
- If s is requesting new slots for an already set up connection, the NC selects the wavelength channel used for this connection.
- On the selected wavelength channel, the NC counts the available slots, excluding among free slots those in which either the source terminal or the destination terminal are busy (possibly for transmitting/receiving data associated to different connections).

- If the amount of free slots resulting from the previous operation is less than the terminal request, the NC looks for another wavelength channel (only for a new connection request).
- If none of the wavelength channels already used for the communication from PON S to PON D is usable, the request is not accepted.
- If the number of free slots matches the terminal request, the NC selects the first available slots in the selected channel (first-fit allocation).
- In the case of datagram request, if the number of available free slots is less than the number of data slots required by the end-terminal, the NC could also decide to operate a partial resources allocation: a threshold can be defined for this goal.

Fig. 3 provides a block diagram of the proposed scheduling algorithm.

For connection oriented (or persistent) requests, resources are allocated until one of the two involved terminals signals that it wants either to release some slots for this connection or to close it. For datagram (or non-persistent) requests, we assume that resources are allocated only in a single frame; in the next frame the NC will release all the resources previously allocated to datagram traffic.

Logical topology design. The logical topology design aims at obtaining an efficient allocation of available channels among PONs. The idea is quite simple: in order to reduce the slot allocation failure rate, the NC can "re-assign" channels: a channel showing a number of free slots above a given threshold L can be freed, obviously re-routing connections previously assigned to that channel; free channels can be assigned to pairs of PONs experiencing a number of free slots below a given threshold H. We compute the load on each channel to decide when and whether or not to allocate/deallocate channels.

In every frame, a pair of PONs (a source PON S and a destination PON D) is selected using a round-robin algorithm. We compute L_{SD}, the number of free slots on all the channels from PON S to PON D. If L_{SD} is larger than threshold L, we try to release a single channel, choosing the least loaded channel. If L_{SD} is below threshold $H < L$, we try to assign another channel to the pair of PONs, provided that at least a free channel is available.

In the process of channel release, we try to release the least loaded channel. This implies a reallocation of all used slots in the selected channel. Only if the process is successful, i.e., all the slots can be allocated

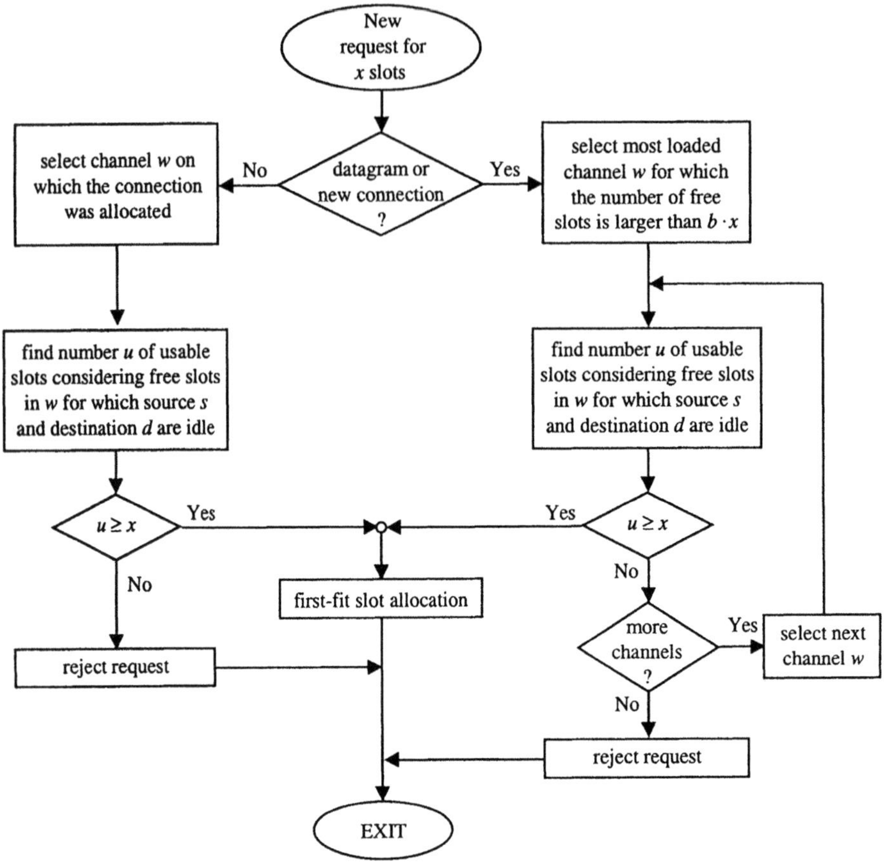

Figure 3 Block diagram for the scheduling algorithm

on other channels, the selected channel can be freed. We perform a complete search on all other channels allocated to the pair of PONs in order to reallocate every single slot, starting from the "wired" channel and continuing on all the other channels in the ordered list. If the reallocation process is successful, the channel can be freed and added to a list of available channels. Otherwise, if even a single slot is not successfully reassigned, the channel cannot be released. Note that, to be more precise, releasing a channel means releasing a wavelength on the transmitter side, and a wavelength on the receiver side.

When trying to add a channel we consider only a single channel addition between the PONs selected according to the round robin scheme. We first look for an available wavelength converter, i.e., a wavelength converter that is not used by both PON S and PON D; if found, we choose the wavelength that should be used at the transmitter and at the receiver, and we logically add the channel to the pair of PONs.

If an available converter is not directly found, we check whether an available wavelength at both the transmitter and the receiver exists. If this is true, we know by Slepian-Duguid theorem [7] that we can rearrange the logical topology so that at the end of the process we obtain an available wavelength converter. Paull's algorithm [7] can be used to obtain the new logical topology, i.e., the new wavelength assignment. The latter algorithm is based on an iterative process that, after an upper bounded number of wavelength reassignments, guarantees the availability of a wavelength converter among the two PONs.

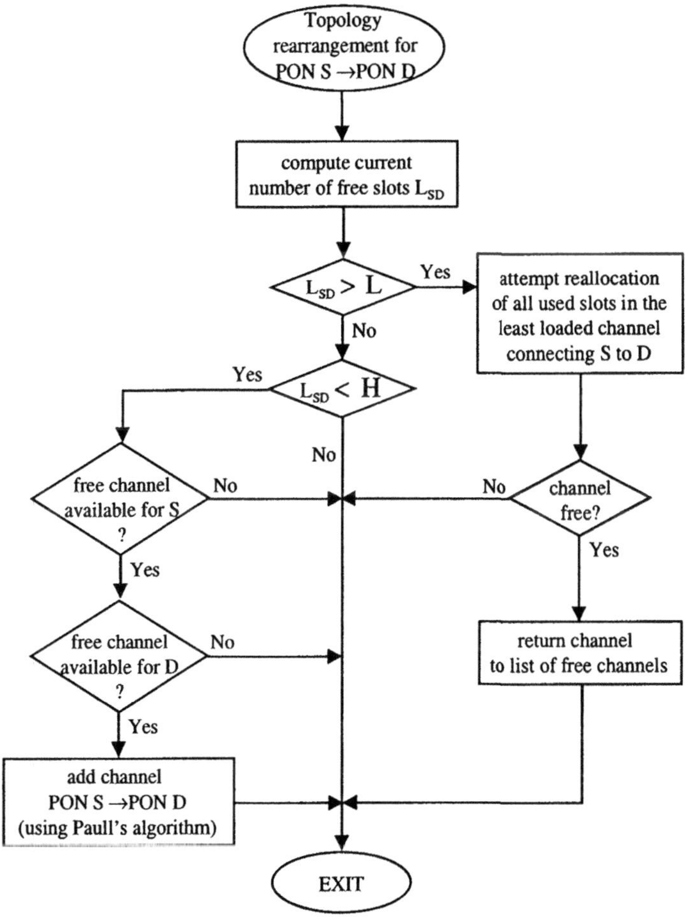

Figure 4 Block diagram for the logical topology design algorithm

Fig. 4 provides a block diagram of the proposed logical topology design algorithm.

Note that each reassignment made while running the algorithm implies updating a significant amount of data structures, which are not reported here for the sake of conciseness. A more detailed description of the algorithms and of the data structures used at the Network Controller can be found in [6].

4. SIGNALING PROTOCOL IN SONATA

End-terminals in SONATA must reserve slots at the centralized Network Controller. Reservation is achieved through the exchange of signaling messages between end-terminals and the NC: N_c (one or more) wavelength channels are available in each PON for signaling. Upstream signaling requires a medium access control strategy, due to the many contending terminals.

A detailed design of a signaling protocol for SONATA was performed in [5]. Signaling messages belong to the following classes:

- slot allocation request (upstream)
- slot de-allocation (upstream)
- slot allocation confirmation (downstream)
- slot de-allocation confirmation (downstream)
- slot allocation rearrangement, either in time or in wavelength, or both (downstream)
- wavelength converter configuration (downstream)

The format of these signaling messages is not reported here because of space limitations.

The use of connection identifiers to label different connection-oriented information flows at each end-terminal was proposed to ease control procedures, and to permit multiple connections between the same pair of end-terminals.

It has been shown that a fixed TDMA upstream access is a viable solution if slots can be partitioned in sub-slots on signaling channels. Indeed, by sub-slotting it is possible to obtain 50,000 minislots in each frame (using 50 sub-slots per slot). With $N_c = 1$, up to 20,000,000 upstream signaling messages can reach the NC in one frame.

For what regards downstream signaling, with the proposed signaling messages, up to at most few thousand allocation requests can be confirmed in one slot. This means that, with $K_c = 1$, up to a few million slot allocation requests can be confirmed in one frame.

This evaluation of the bandwidth required for signaling is an important outcome of the design of the signaling protocol. Obviously, more bandwidth is required for downstream signaling than for upstream signaling. It is also easy to see that there is an almost mandatory need for a dedicated fixed receiver at each terminal for signaling (in addition to the tunable transceiver for data).

The optimal control of SONATA, in which wavelength converters can be reconfigured slot by slot, can be shown to require a large amount of signaling bandwidth (in addition to a very large complexity at the NC): approximately 1.5% of the bandwidth available for data transfers would have to be made available to downstream signaling in this case. Note that the proposed approach to network control assumes to let existing allocations unaffected by new allocations (this is often called "transparency" property in the technical literature), thereby significantly reducing the amount of downstream signaling.

The signaling bandwidth required for slot/topology rearrangements is instead negligible with respect to the bandwidth required for slot allocation/de-allocation (see [5, 6]).

5. SIMULATION OF THE PROPOSED RESOURCE ALLOCATION ALGORITHMS

In this section we study by simulation the resource allocation algorithms, in terms of their speed of convergence to the optimal logical topology, their stability, and their capacity. By optimal topology we mean the topology that maximizes the network throughput in a given traffic scenario.

We first consider a small system, with 3 PONs, $N_t = 150$ end-terminals, i.e, 50 users per PON, frame length $F = 100$, $N_d = 3$ dummy ports connected to a bank of 3 wavelength converter arrays and 6 wavelengths per fiber. We modeled both network controller resource allocation algorithm, i.e., logical topology design and scheduling, under uniform traffic pattern. We discuss the speed of convergence of the algorithm to the optimal logical topology as a function of the system load. Note that in this case the optimal topology is obtained by allocating the same number (equal to 2) of channels to each PON pair.

At time 0, all 9 additional channels provided by the programmable portion of the network are allocated between PONs connected to input and output ports with the same index, i.e., 4 channels are available between source PON I and destination PON I and 1 (wired) channel is available between PON I and all other PONs. Thus, the topological distance is 12, since each PON has 2 additional channels allocated to reach its end-terminals, and one channel missing towards the other two paths. Note that this is a logical topology that maximizes the topological distance[1] from the optimal topology under uniform traffic.

We show in Fig. 5 the topological distance of the current topology from the optimal topology as a function of time (one unit refer to a time frame), for variable loads. For medium-high loads ($\lambda = 0.9$), the system converges rapidly to the optimal topology; once the optimal topology is reached, due to traffic fluctuations, the algorithm sometimes modifies the logical topology, mainly releasing free channels; the topological distance from the optimum remains small. Results for smaller loads not reported for the sake of conciseness show a very similar behavior.

For moderate overload ($\lambda = 1.4$) the optimal topology is reached quite quickly; once all the channels are correctly allocated to pair of PONs, the system does not modify the topology since no channels become available at any time.

For sustained overload ($\lambda = 2.4$) the optimal topology is never reached; similarly to the previous scenario, once all the channels are allocated to pair of PONs, the system is not able to modify this topology since no channel becomes available at any time. However, the system does not converge in general to the optimal topology but to a topology that depends on the initial stochastic behavior of call requests.

To analyze the capacity provided by the proposed algorithms, we modeled again a 3 PON network, but with 10 users per PON and 10 slots per frame, subject to telephone traffic. Each call requires exactly one slot. A Poisson process models call arrivals, and call holding times are exponentially distributed with average 18.000 frames. Part of the traffic is uniformly distributed among PONs; the remaining traffic is randomly sent to terminals belonging to the PON to which the originating terminal belongs. The percentage of local traffic is taken as a parameter. The offered load is normalized with respect to the total number of slots available per frame, and the plotted throughput is the fraction of calls that are successfully routed through the SONATA network.

Figure 6 plots the network throughput under different locality factors; as a comparison, we added also a curve relative to the optimal scheduler that is able to route all calls up to the maximum number of available

[1]We define as topological distance between two topologies the following quantity:

$$\sum_S \sum_D |C^1_{SD} - C^2_{SD}|$$

where indexes S and D span over PONs, C^1_{SD} and C^2_{SD} represent the number of channels allocated from PON S to PON D in the first and in the second logical topology respectively.

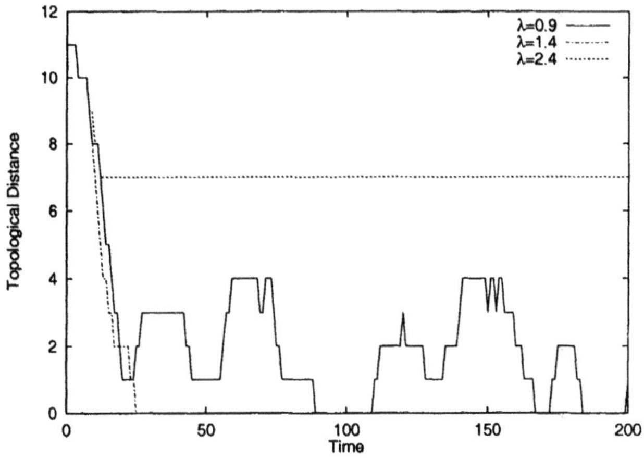

Figure 5 Convergence of the current logical topology to the reference topology

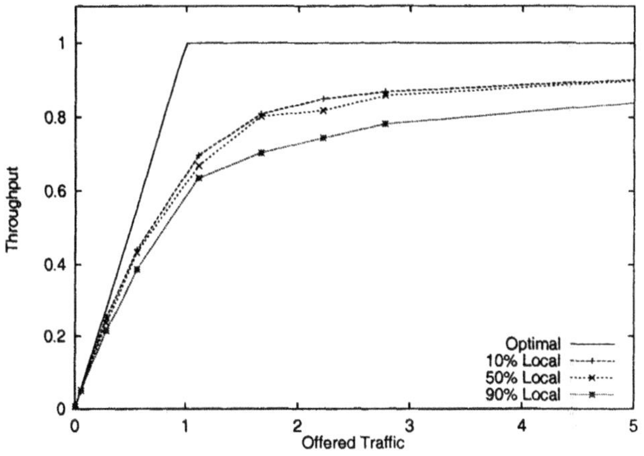

Figure 6 Throughput versus the offered load, with different traffic pattern.

slots, possibly completely reconfiguring the network at each frame, regardless of transmitter and receiver contentions. The performance of the proposed algorithms is not too far from the optimal behavior. When the amount of traffic locality is larger, a lower utilization of direct channels (which cannot be re-allocated according to traffic needs), hence a lower throughput, is experienced, as expected.

These results show that the network control algorithms proposed by the SONATA project lead to acceptable exploitation of the very large available bandwidth with a reasonable complexity.

6. CONCLUSIONS

The paper described the network architecture and provided a performance analysis of a passive optical network named SONATA, which has been proposed and demonstrated in the context of the European Union ACTS Program. SONATA aims at avoiding the need for large and fast switching electronic nodes in a high-speed nation-wide network. To reach this goal, the network structure and the layer architecture within the network have been drastically simplified: the transport infrastructure consists of a single-layer of end-to-end optical connections. End-terminals fully exploit time and wavelength agility to exchange packetized information in the multiple-access network.

The control of the network is centralized at a network control device, whose primary goal is to assign time/wavelength resources to terminals in a way such that conflicts among transmitters and receivers are avoided. The paper focussed on the algorithms that must be executed at the Network Controller to solve the resource allocation problem.

Since the resource allocation problem at the Network Controller is in general NP-hard, we proposed in this paper simple heuristic algorithms that divide the solution into scheduling and logical topology design subproblems, aiming at a sub-optimal approach with a limited complexity.

An initial performance evaluation of the considered algorithms, based on simulation, was presented for a small scale system.

References

[1] B. Mukherjee, "WDM-Based Local Lightwave Networks Part I: Single-Hop Systems", *IEEE Network*, vol.6, pp.12-27, May 1992.

[2] D. Banerjee, B. Mukherjee, "A Practical Approach for Routing and Wavelength Assignment in Large Wavelength-Routed Optical Networks", *IEEE Journal on Selected Areas in Communications,* vol. 14, n. 5, pp. 903-908, June 1996.

[3] R. Ramaswami, K. N. Sivarajan, "Design of Logical Topologies for Wavelength Routed Optical Networks," *IEEE Journal on Selected Areas in Communications,* vol. 14, n. 5, pp. 840-851, June 1996.

[4] M. Hill, N. P. Caponio, F. Neri, R. Sabella, "Single Layer Optical Platform Based on WDM/TDM Multiple Access for Large Scale "Switch-less" Networks", to be published in *European Transactions on Telecommunications,* Special Issue on WDM Networks, January 2000.

[5] A. Bianco, E. Leonardi, M. Mellia, M. Motisi, F. Neri, *Specification of Signaling Functions in SONATA*, ACTS Project SONATA AC-351, Working Document, Politecnico di Torino, 1999.

[6] A. Bianco, E. Leonardi, M. Motisi, F. Neri, *Specification of Slot Allocation and Topology Design Protocols,* ACTS Project SONATA AC-351, Working Document, Politecnico di Torino, 1999.

[7] J. Hui, *Switching and Traffic Theory for Integrated Broadband Networks*, Kluwer Academic Publisher, 1990.

Hybrid Fiber Radio Systems: Features and Performance

Roberto Sabella
Ericsson Telecomunicazioni, R&D-GPC, Roma, Italy; E-mail:
roberto.sabella@tei.ericsson.se; fax +39 6 72597439; tel. +39 6 72597442; Via An agnina 203, 00040, Rome, Italy.

Key words: Fiber-to-the Area Systems (FTTA), Hybrid Fiber-Radio Systems, Sub-Carrier Multiplexing, Access Network Architecture.

Abstract: The paper summarizes basic system design issues for an optical fiber feeder in either mobile radio or fixed radio access systems. The features of the introduction of the optical feeder are discussed, and the performance evaluated in some relevant cases, are reported.

1. INTRODUCTION

The low loss and large band features of the optical fibers have led to the realization and the installation of a variety of advanced communications systems throughout the world. At the same time, mobile communication systems, which allow people to communicate at anytime, anyplace and with anybody, are more and more diffused. Furthermore, the radio resource is going to be even more exploited to distribute broadband services to both business and residential customers, by using millimeter wave band. Fixed radio access systems are representing a valid alternative to realize access systems without the need of costly cable infrastructures.

The combination of fiber optic and radio technologies allows the realization of hybrid fiber radio (HFR) systems (even known as FTTA: fiber-to-the-antenna), which have an important advantage: easy and flexible connection of cells/microcells and, particularly, greatly simplified radio base stations. In addition, these systems can be upgraded very quickly, because there is no need to replace existing base station equipment.

Research in this field started several years ago (see as an example the whole special issue [1]), focused on microcellular systems. In fact, it is well known that microcellular systems, composed of small size cells with several hundred meter diameter communication range, promise effective frequency spectrum utilization. However, these systems require a large number of base stations. Therefore, especially for these systems, miniaturization of base stations is an important issue in realizing microcellular systems.

The HFR approach allows the miniaturization and a significant simplification of the base station architecture. Section 2 examines the HFR approach in a mobile case, while section 3 considers the FTTA case for wireless broadband systems. Finally, section 4 reports a discussion about the performance of these systems, and some general conclusions.

2. HFR SYSTEM FOR MOBILE COMMUNICATIONS

The hybrid approach was originally introduced for being applied in microcellular communication system [2]-[4].

In such systems, service areas are divided into many zones named "microcell" having radius of some hundred meters. In fact, since the base station's antennas are located about 10-15 m above the ground, they can only cover small area. This system approach allows frequency reuse, and can accommodate a large numbers of subscribers. On the other hand, this system has the disadvantages that a large number of base stations (BS) are required because of the large number of microcells required to cover a service area. Furthermore, personal stations frequently move across microcells, so channel assignment and hand-off procedure is much complicated.

To solve this problem, it was proposed to connect microcell BS with a microcell control station (CS) using an optical fiber link, in which the signal obtained by frequency division multiplexing of the telephone channels directly modulates a laser source (SCM: sub-carrier multiplexing) [2]-[4]. In this kind of system, the BS only converts the radio frequency signal to an optical signal: the required equipment for BS is remarkably simplified compared with the conventional system. The BS simplicity is an important feature of the system, because many BS would be typically required and, therefore, they should be economical and trouble-free in remote locations. Furthermore, channel assignment and hand-off procedure are centralized. The layout of the microcellular system is identical to the existing cellular radio one except fibers and base stations have been inserted between the mobile units and the central stations.

Intensity modulated optical signal is transmitted over optical fiber to microcell CS, where optical signal is directly detected using a p-i-n photodiode, and a bandpass filter is used to provide RF channel signal. After filtering, each RF signal is demodulated and connected to the public switched telephone network. The system configuration is schematically reported in figure 1.

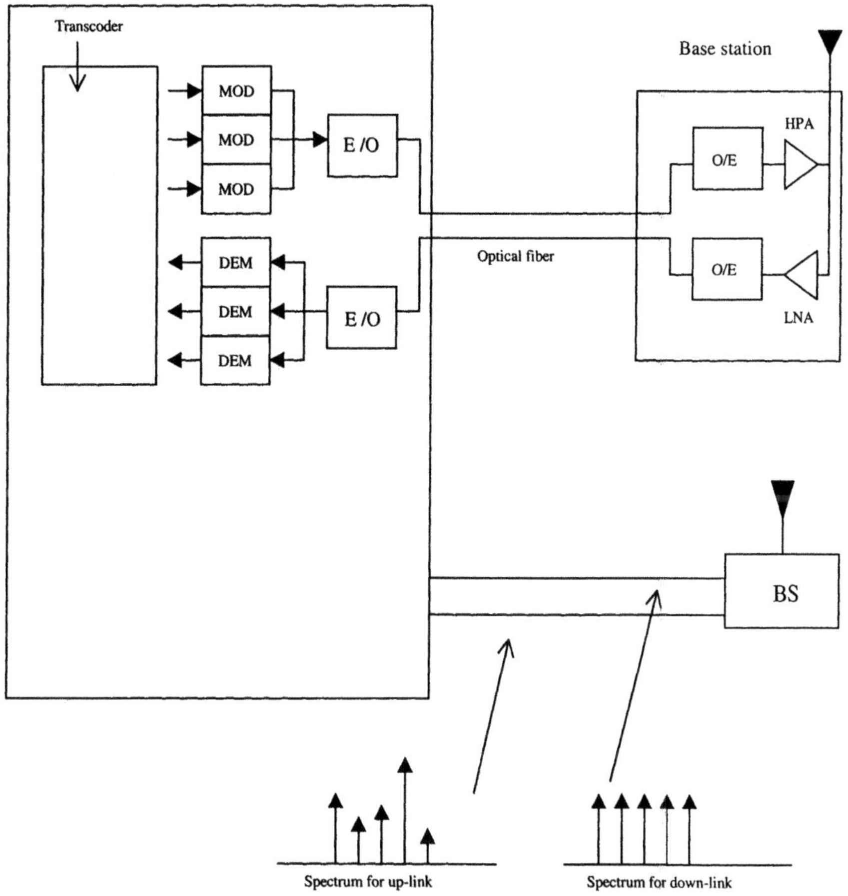

Figure 1. Basic configuration of FTTA system

For mobile radio applications, the optical feeder is required to have low noise and low-distortion characteristics. This requirement is particularly stringent for the up link feeder, since the radio signal received at the BS fluctuates greatly due to fading, shadowing and changes in distance between the micro BS and the mobile stations.

Although laser diode characteristics in analog modulation have been improved with respect to the past (e.g. in 90s), the feeder dynamic range is still limited by laser diode noise and non-linearity. So, using SCM technique, optical source non-linearity must be taken into account. In fact the non-linear characteristic of the laser diode causes several inter-channel interference, especially when many sub-carriers are transmitted from a single source, that is, for high total modulation indices. Furthermore, even if the laser characteristic was linear above its threshold current, signal distortion occurs also because the laser intensity is occasionally clipped by the large negative excursions of the modulating current. This phenomenon called clipping, gives rise to impulsive noise that deteriorates the system performance [5]. Optical source non-linearity is sketched in figure 2.

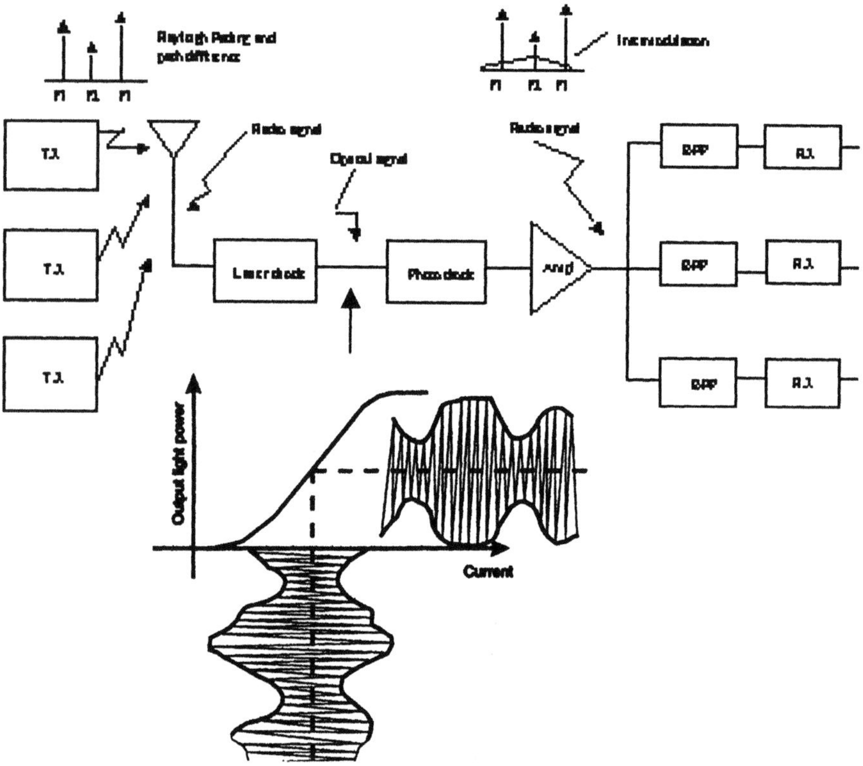

Figure 2. System degradation due to third-order intermodulation terms and clipping noise

The model that allows the evaluation of the transmission performance of the systems taking into account all these effects was reported in [6].

3. FTTA SYSTEM FOR BROADBAND COMMUNICATIONS

The employment of the optical feeder in broadband wireless systems is a very attractive option.

Before the description of the FTTA scheme considered in this paper it is worth while briefly mentioning the enabling technologies. The fiber optic technologies for microwave and millimeter-wave transmission are reviewed in [7], as well as different link architecture for RF signal transmission (i.e. directly modulation of the laser, external modulation, etc.). Furthermore, that paper discussed MMIC techniques for optical receivers, and stated that HEMT and HBT devices are very attractive in the realization of compact and cost effective optical receivers.

The enabling technologies were also discussed in a more recent paper [8]. In particular, the latter paper discussed three different approaches for linking the optical feeder with the radio link:

1. Baseband transport between the CS and the BS. This is the state of the art; in this case commercial equipment's can be used. However, when using this concept, the BS concentrates all the equipment's that depends on the physical medium of the final drop (e.g. coding, error correction, modulation and upconversion to IF, upconversion to RF, filtering and amplification).

2. Intermediate frequency (IF) transport. It makes possible to use BSs with fewer functions. This approach requires an optical IF transceiver, up/down conversion between the intermediate and radio frequencies, power amplification on the downlink, and a low noise front end on the uplink. All other functions, like line termination of the digital segment in the core network, mux/demux functions and router to map partial data streams to radio channels, access management for the shared radio medium, modems for signal adaptation to the radio channel, and up/down conversion to the IF, are performed in the CS.

3. Optical mm-wave RF transport. This is an advanced and attractive solution. Here the BS consists of an optical RF transceiver followed by a stage providing power amplification on the downlink and low noise amplification on the uplink. Promising techniques for the practical realization of mm-transport are based on a self heterodyning concept [9], and on the external modulation of the laser source by means of electro-optical modulators working at mm-waves [10].

In this paper we basically refer to the second approach (IF transport), which is synthetically depicted in fig. 3 [11].

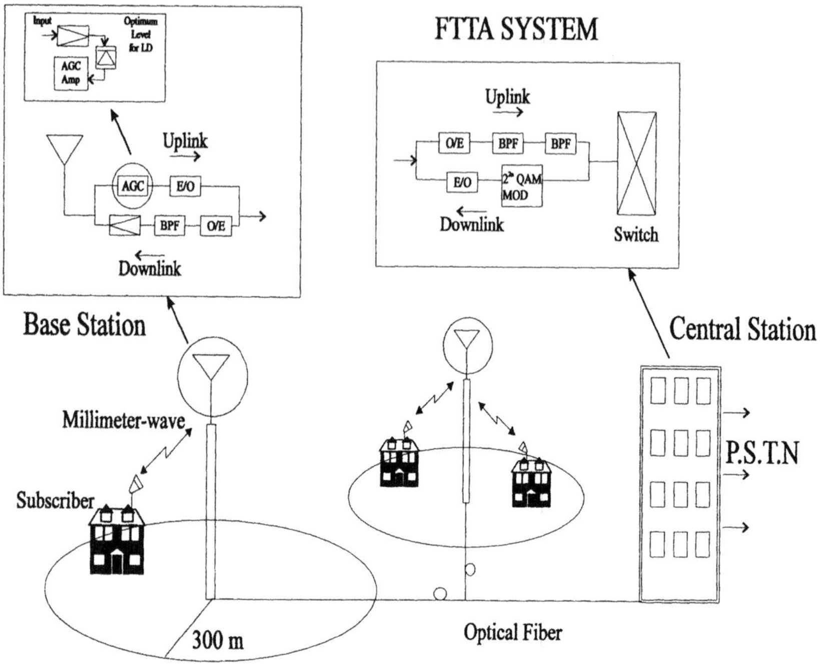

Figure 3. The concept of a FTTA system. P.S.T.N.:Public Switched Telecomunication Network; O/E or E/O: optoelectronic or electro/optic converter; BPF: bandpass filter; MOD: modulator;DEM: demodulator; AGC: automatic gain control circuit

In this system each zone radius can be of the order of hundred meters (between 300 m and 1 km in the considered cases). In the uplink, base station (BS) receives MW signal from air interface (comb of frequency division multiplexed signals) and converts it in an optical signal. Such a sub-carrier multiplexed (SCM) signal in then transmitted to the control station through optical fiber. Therewith, the control station (CS) converts back the SCM signal into the electrical domain, performs the demultiplexing operation, assigns each signal to a channel, and connects it to public switched networks. The downlink works in a similar way. It is also assumed that the double-side bandwidth per carrier is in the range 20 - 150 MHz, to transmit future broad band signals such as 150 or 600 Mbps, with the use of modulation schemes with good spectral efficiency, such as QPSK (Quadrature Phase Shift Keying) or 22n QAM (Quadrature Amplitude Modulation).

The main features of the considered approach, with respect to traditional microcellular systems, can be summarized as follows:

1. the BS is simplified, since it does not perform neither modulation/demodulation operations nor control functions (they are indeed achieved in the CS), but it simply executes amplification, frequency

up/down conversion (usually named RF/IF conversion), automatic gain control (AGC) functions and E/O (O/E) conversion;

2. since the BS is simplified the cost is lowered;
3. the optical feeder gives high flexibility for changes in modulation schemes because it is transparent for all radio signals;
4. the FTTA approach represents an economically viable solution because it could reuse the existing network terminations and the terminals for coax and hybrid fiber coax (HFC) systems.

The efficiency of the operations made in the BS (amplification, E/O conversion etc.) is not critical to assess the feasibility of the present system. Therefore, in this paper, we assume that ideal devices do such operations. In fact, if such non-ideal efficiencies were considered, a further small degradation of the carrier-to-noise ratio would come out. Anyway, this degradation does not represent a real limiting factor for the systems performance as a whole.

The transmission performance of the FTTA system is influenced by several factors. As far as the radio link is concerned, the outage probability, due to the rainfall attenuation, is most important one. While, concerning the optical link several issues impact the performance, such as the effect of the resonant frequency in the laser intensity modulation, the non-linear characteristics of the laser, in particular clipping, and the interplay of laser chirping with fiber dispersion.

The model reported in [12], based on a physical model of the laser dynamics, allows the evaluation of the transmission performance, by taking into account, simultaneously, all the effects influencing the performance.

4. PERFORMANCE OF FTTA SYSTEMS

To understand the features and the limits of the FTTA approach in either mobile communication systems or in broadband wireless systems, the following two sections report some relevant results for both the cases.

4.1 Performance of FTTA systems for mobile communications

To evaluate the performance transmission of a FTTA system in term of CNR, it has been realized a simulator that generates N (where N is the number of RF carriers that modulate the laser) Rayleigh distributed random variables. The generated set of variables is utilized to calculate the CNR, relative to the statistic mean of the optical modulation indices. To obtain results relative to a range of OMI values of almost 40 dB it is necessary to repeat the previous operations a high number of times. Furthermore, the simulator evaluates the CNR for non-faded case, that is, supposing that all the different carriers reaching the antenna have the same optical modulation index. To evaluate the performances of a system affected by multipath fading and optical non-linear distortions, (clipping and static intermodulation), some simulations have been done. The analysis refers, first, to a case in which 3 carriers reach the antenna and the transmission parameter reported in table 1 have been utilized. The average CNR of the central carrier is evaluated, because the carrier in the middle of the frequency band suffers from most severe non-linear distortion [13].

Figure 4 depicts the average CNR versus the average OMI. The results refer to a case in which clipping is not taken into account. This is not a practical case, but this kind of simulation allowed us the comparison between our simulation results and the ones existing in literature [13].

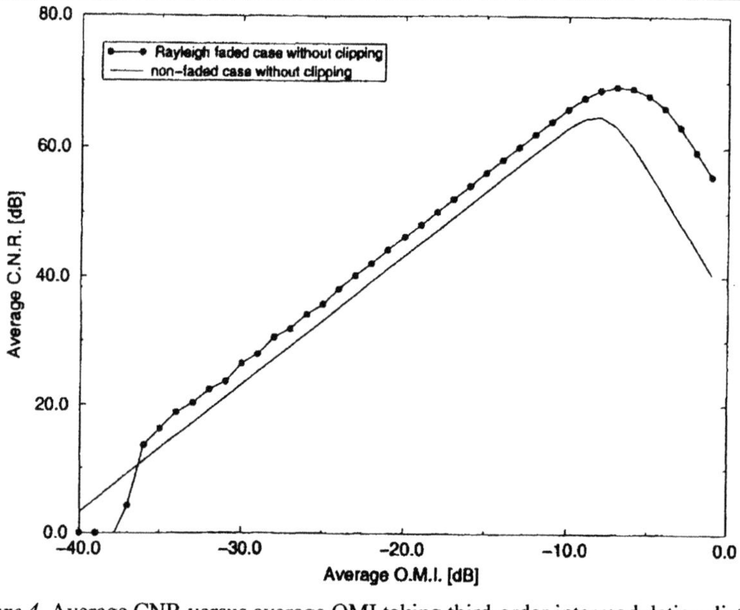

Figure 4. Average CNR versus average OMI taking third-order intermodulation distorsion into account. Comparison between Rayleigh faded and nonfaded case. Three carriers per microcell are considered

In particular in this figure the average CNR both for Rayleigh faded and non faded case is reported. It can be noticed that when the signals suffer from fading, average CNR slightly improves. The reason of this is explained as follows: the third order intermodulation distortion is the product of three different RF signals, and the effects of this are significant only if the signal levels of the three different carriers are simultaneously large, but such a case is rather unusual.

In figure 5 the comparison between faded cases with and without clipping effects is reported. For low optical modulation indices, the same results are obtained: this means that clipping noise is not dominant and fading multipath increases the average CNR. On the contrary, for high OMI, when clipping noise power becomes dominant, the average CNR of the signal affected by clipping decreases. It is worth noting that clipping occurs when the modulating current exceeds the threshold current: this effect is dominant if just one of the RF carriers has a signal level that can be clipped by the laser.

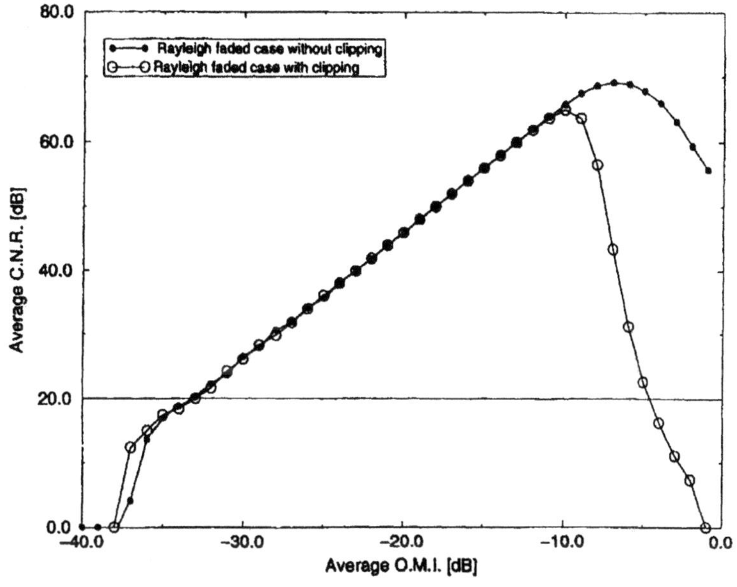

Figure 5. Average CNR versus average OMI. Comparison between Rayleigh faded case with and without clipping. Three carriers per microcell are considered

The simulations are repeated intensifying the number of carriers per microcell. In particular, different cases are considered: 3,5,7,10,13,17 and 20 carriers per microcell. Figure 6 shows the results obtained taking clipping into account and considering the Rayleigh fading. It is worth noting that the maximum OMI that can be employed to obtain a CNR greater than 20 dB diminishes if the number of carriers increase: this is due to laser clipping effect. Furthermore, as is better shown in figure 7, the maximum value of CNR drops with the growth of the number of carriers. In fact, when the number of carriers that modulate the laser arises, the noise power due to static intermodulation, that is the IM3 power, arises too, and, consequently the carrier to total noise decreases.

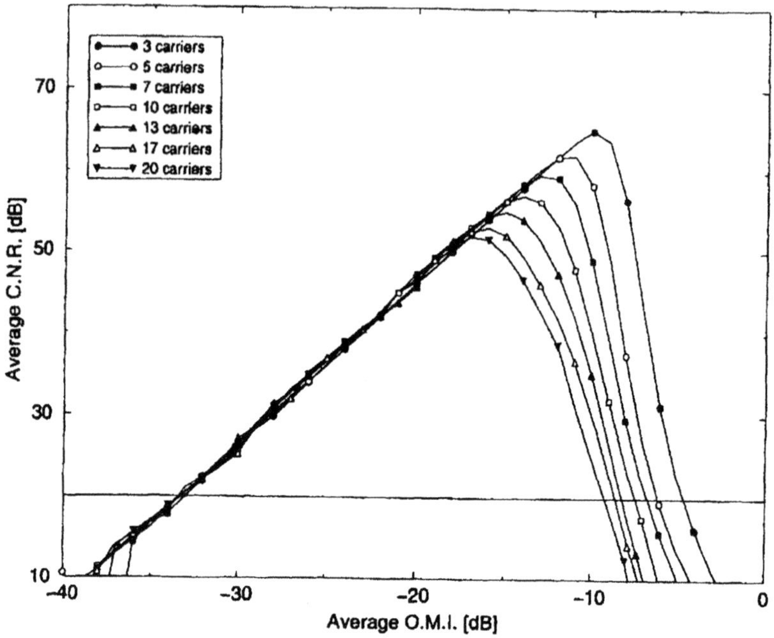

Figure 6. Average CNR versus average OMI for different carrier numbers per microcell

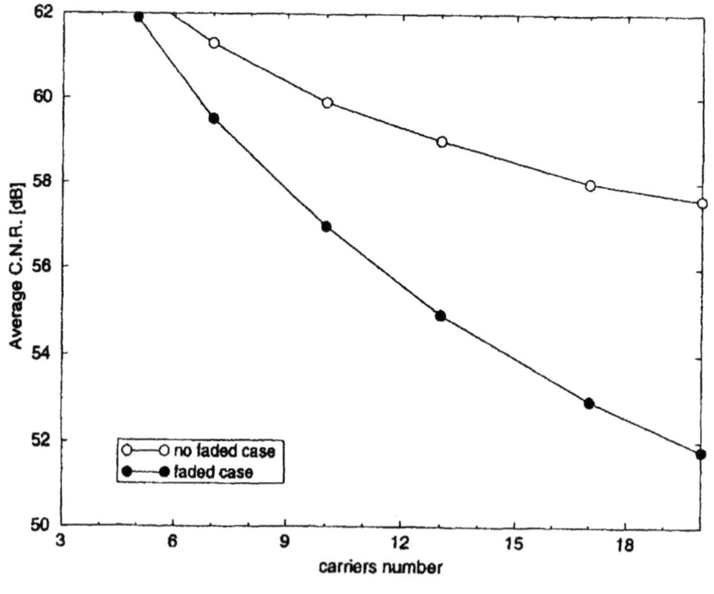

Figure 7. Maximum CNR versus carrier numbers per microcell. Both Rayleigh faded and nonfaded cases are reported

It has already been underlined that, when RF carriers reach the E/O converter, they have experienced different degrees of propagation loss, multipath fading loss and shadowing loss. This implies that a wide range of power level variation may exist in the received carriers. Now the range of this power level variation is considered, since it has important implication in the dynamic range requirement of an optical fiber link and in particular of the E/O device.

Furthermore, a radio/optical link power budget is reported, realized employing the link parameters reported in table 1. In particular, the maximum and minimum LD input carrier intensity (PO) that can modulate a laser and that allows to obtain a fixed CNR is evaluated. These current values are directly related to the optical modulation index by the expression

$$m = \sqrt{\frac{2 \cdot P_{in}}{R_{LD}} \cdot \frac{\eta_{LD}}{P_o}} \qquad (1)$$

To study the feasibility of the link, these values are compared with the maximum and minimum RF power that reach the BS antenna. The received power (Pr) can be evaluated using the following expression:

$$P_r = S \cdot P_t \cdot \left(\frac{\lambda}{4 \cdot \pi \cdot R} \right)^2 \qquad (2)$$

where P_t is the power transmitted by the mobile station, λ is the wavelength of the frequency of interest, R is the distance between a handset and a base station and S takes into account the shadowing effects (S=1 in line-of-sight (L.O.S) propagation). It is assumed that the maximum RF power occurs when a mobile station is placed in L.O.S with the BS station, and it is next to it (this means that the distance between the transmitting and the receiving antennas is about 10-15 m). While the minimum power occurs when a mobile station is placed and the edge of the cell (that is R is equal to the cell radius). In this case the received power is affected by path loss and shadowing. In the calculation, the parameters reported in [14] have been used, where the maximum signal attenuation due to shadowing is set to 30 dB.

The comparison between the maximum and minimum RF received power and the maximum and minimum LD input carrier intensity is necessary to estimate if there is an optimum value of the low noise amplifier (LNA) gain that allows the realization of the link. In particular the radio/optic link is feasible, only if the difference between the maximum power values is greater than the difference between the minimum ones. This condition assures that, an optimum LNA gain value, that permits to obtain a CNR greater than the requested one, exists.

Table 1 reports the results obtained in different case, according to the minimum requested CNR (15 or 20 dB in the considered cases); and the microcell radius (200 or 300 m). To reduce the dynamic range of the received signals, and to guarantee the link accomplishment, different techniques have been proposed: antenna pattern discrimination, automatic power control at the mobile station or macrodiversity. The latter technique, however, that employ the cell splitting concept, don't assure the base station equipment simplification. In fact every BS must be equipped with three 120° directional antennas and with the adequate electronic devices [15], [4].

	Microcell Radius = 200m		Microcell Radius = 300m	
	CNR = 15 dB	CNR = 20 dB	CNR = 15 dB	CNR = 20 dB
3carriers	Possible with G = 21.11 dB	Possible with G = 25.51 dB	Possible with G = 24.63 dB	Possible with G = 29.03 dB
5 carriers	Possible with G = 20.31 dB	Possible with G = 25.11 dB	Possible with G = 23.83 dB	Not possible
7 carriers	Possible with G = 19.91 dB	Possible with G = 25.01 dB	Possible with G = 23.43 dB	Not possible
10 carriers	Possible with G = 19.71 dB	Not possible	Possible with G = 23.25 dB	Not possible
13 carriers	Possible with G = 19.51 dB	Not possible	Not possible	Not possible
17 carriers	Possible with G = 19.49 dB	Not possible	Not possible	Not possible
20 carriers	Possible with G = 18.89 dB	Not possible	Not possible	Not possible

TABLE 1 Microcell Link Power Budget

4.2 Performance of FTTA systems for wireless broadband communications

The performance of the optical link has been reported in terms of the CNR versus the optical modulation index (OMI) of the single channel. It is worth noting that such an analysis is similar to the one we reported in [16]. However, the total bandwidth, in the cases treated in the present paper, is quite different than in the case of CATV signal distribution: we consider here, in some cases, a total bandwidth of the order of the GHz, that is quite larger than the one considered for CATV systems. In fact, we expect a greater impact of the chirping/dispersion impairment.

In fact, figures 8 - 10 show the transmission performance of the system, in terms of the CNR versus the optical modulation index (OMI) of the single channel, varying several system parameters.

A general observation of all those figures, which is worth remarking, is that the increasing of the signal power, directly related with the OMI, produce an increasing of the CNR, until non-linear distortions become significant, so causing an increasing of the NLD contribution to the noise power. This explains why there is an optimum value of the OMI, which allows the maximum CNR to be reached.

The impact of the number of SCM carriers on the performance is depicted in figure 8, where the performance are reported for different values of the number of carriers, assuming a double-side bandwidth of 150 MHz for any single channel. In fact, it can be noticed that the CNR decreases with the increasing of the number of carriers. Basically, two reasons can be individuated for this: i) the overall bandwidth of the SCM signal increases and, consequently, the noise contributions grow accordingly; ii) the probability of clipping events (e.g. the clipping induced total distortion) increases when there are more carriers [16].

To investigate the impact of the single channel bandwidth on the system performance, figure 9 shows the CNR versus the OMI for three different values of the channel bandwidth, assuming the transmission of 20 SCM channels. It is evident as the increasing of the bandwidth leads to an increasing of the noise contributions. Moreover, the increasing of the bandwidth, as a whole, could generate further distortion contributions due to both chirping/dispersion and the effect of the resonance frequency of the laser.

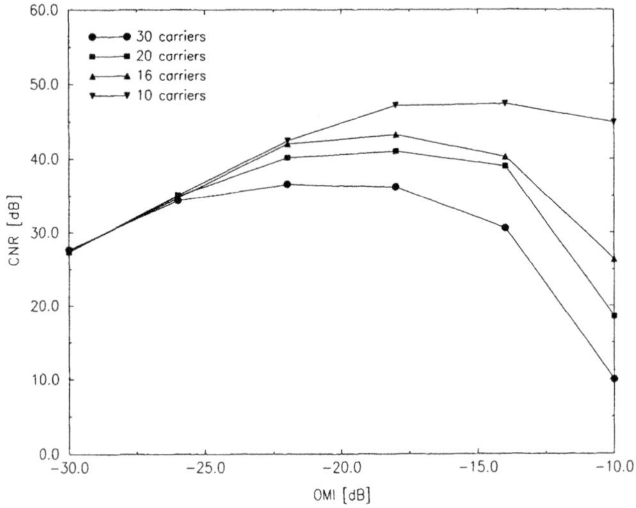

Figure 8. CNR versus the OMI, for different numbers of carriers; the RF comb starts from 200 MHz

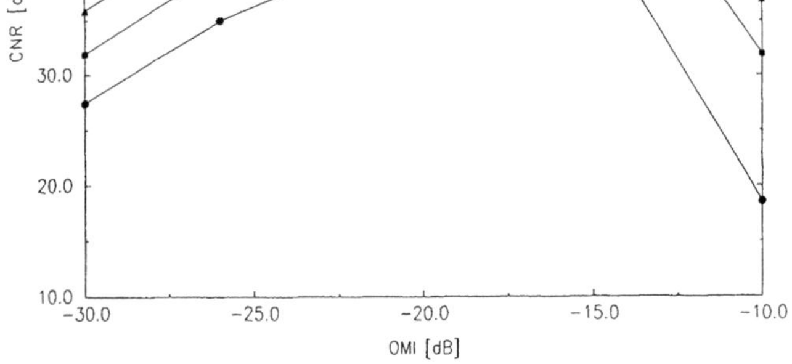

Figure 9. CNR versus the OMI, for different values of the channel bandwidth; the RF comb starts from 200 MHz

In order to evaluate the effect of source chirping (in commingling with fiber dispersion) and of the resonance frequency of the laser on the system performance, it is useful to observe figure 10. Here it is assumed to transmit 20 carriers. It is also presumed a fiber optic link length of 30 km. Such a feeder length is realistic, if we consider that a control station links a certain number of base stations located in a geographical area. The MW signal, consisting of channels' comb, is received from the antenna, and is down shifted near the baseband (200 MHz) in one case, and to 1 GHz in the other case. In both the cases the performance are evaluated either considering the chirping effect or not.

By comparing the curves relating to the comb initial frequencies of 200 MHz and 1 GHz, respectively, it is possible to observe a difference of few dB (3 in the worst case). This is due to the fact that, when the laser operates close to the resonance frequency (the relaxation oscillation frequency), further distortions are introduced. Therefore, the channels' comb has to be placed in order to avoid this effect.

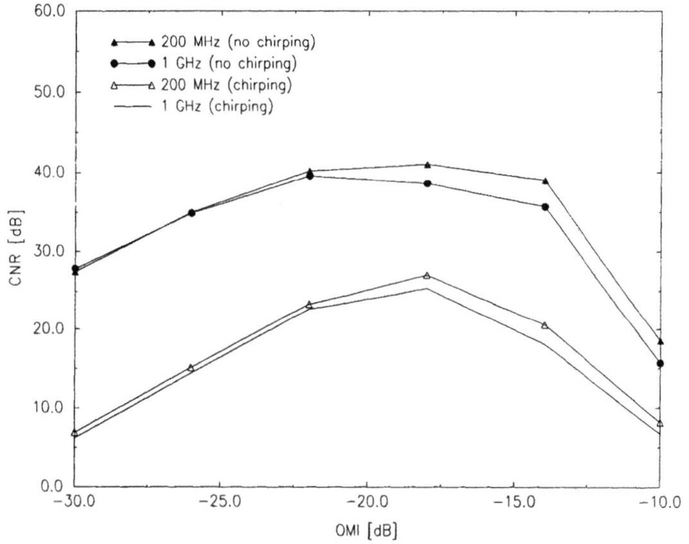

Figure 10. CNR versus the OMI, for different values of the start frequency and showing the chirping effect

As far as the system design is concerned, it is useful to derive some qualitative considerations from the obtained results. For instance, since the required CNR values, which allow an error probability of 10-6 to be reached, for QPSK and for 16 QAM modulation schemes, are 13.8 and 18.8 dB, respectively, we could say that the considered system can transmit up to 30 carriers, each one having a double-side bandwidth of 150 MHz. On the other hand, if the 64 QAM modulation scheme is desired, the required CNR value of 34.8 dB pose restrictions on the number of carriers, and requires a limited range of value of the OMI (with 20 carriers the OMI should be in the range -26 ÷ -13 dB). This means that QPSK or 64 QAM modulations can be favorably adopted for this application.

It is also to be highlighted that the impact of chirping effect is absolutely not negligible since, in case of long fiber optic link in the trunk (e.g. 30 km), it brings about a penalty of more than 13 dB, on the CNR value.

Again, considering the required values of CNR for QPSK and 16 QAM modulation schemes, we could say that, in this case, just a limited range of values for the OMI is admissible, considering a transmission of 20 carriers of 150 MHz bandwidth each.

5. DISCUSSION AND CONCLUSIONS

The application of the hybrid approach in either mobile communication systems or in broadband wireless systems can be successful.

In the analysis of mobile systems, both Rayleigh multipath fading and optical non-linear distortions were taken into account. The results, obtained with several carriers transmitted from a single source, show that, for high optical modulation indices, clipping effect, more than static intermodulations effect, diminishes the average CNR at the photodetector receiver. Therefore, the number of carriers allowed to be used per microcell is drastically limited by clipping noise: to assure a good service quality it is indispensable to regulate the number of carriers per microcell or employ adequate technique to reduce the dynamic range of the received signals.

As far as the wireless broadband systems, the following considerations can be made. The rainfall attenuation is the basic limitation of the performance of the radio link: it introduces an outage probability. A typical and relevant example of FTTA system has been presented and a performance analysis has been achieved. In fact, in the case of small cell radius (300 m) the outage probability was evaluated to be acceptable, while in the case of large cell radius (1 km) the outage probability was too large for the considered applications.

As far as the optical link is concerned, noise contributions, non-linear distortions and the combined effect of chirping/dispersion are the most limiting factors. In particular, optical non-linear distortions are basically due to the threshold characteristic of the laser, which gives rise to clipping effect. Nevertheless, other non-linear contributions, such as the non-linearity of the laser characteristic even above threshold or the saturation effect, are intrinsically taken into account into the physical model used in the present analysis.

The analysis revealed that this kind of system allows the transmission of some tens of broadband carriers, using two relevant modulation schemes: QPSK and 16 QAM.

In general, it is possible to say that a tradeoff exists between the number of carriers and the channel bandwidth. Furthermore, the impact of chirping could be significant.

As a result, the analysis reported in this paper assesses the feasibility of the FTTA approach for realizing broadband services distribution systems. In the design of this type of systems it is necessary to consider the following points.

1. Dimensioning of the radio link, in order to obtain a fixed outage probability.
2. In practical systems automatic gain control circuits are required in the base station to compensate for the rainfall attenuation.
3. The optical link has to be designed taking into account non-linear distortions contribution arising by direct modulation of a single mode laser source, and the chirping/dispersion contribution, especially if the feeder length is some tens of kilometers long.

6. ACKNOWLEDGEMENTS

I wish to thank my student Alessandra Rohr, for her help in preparing this paper.

7. BIBLIOGRAPHY

[1] Special Issue on Fiber-Optic Microcellular Radio Communication System and Their Technologies, IEICE Trans. on Comm., vol. E76, no. 9, pp. 1059-1185, 1993.

[2] K. Morita, H. Ohtsuka, "The new generation of wireless communications based on fiber-radio technologies", IEICE Trans. on Comm., vol. E76, no. 9, pp. 1061-1068, 1993.

[3] J. Namiki, et al., "Optical feeder basic system design for microcellular mobile radio", IEICE Trans. on Comm., vol. E76, no. 9, pp. 1069-1077, 1993.

[4] W.I. Way, "Optical fiber-based microcellular systems: an overview", IEICE Trans. on Comm., vol. E76, no. 9, pp. 1091-1102, 1993.

[5] C.J.Chung, I.Jacobs, "Simulation of the effects of laser clipping on the performance of AM-SCM ligthwave systems", IEEE Photon.Techn.Lett., Vol.3, N°11, pp.1034-1036, 1992.

[6] R.Sabella," Performance evaluation of optical fiber link in microcellular mobile communications systems ,"Microwave and Optical Technology Letters, Vol. 18, no 3, June 1998.

[7] H. Ogawa, " Microwave and millimiter-wave fiber optic technologies for subcarrier trasmission systems," IEICE Trans. Electron., vol. E76-B, no 9, pp. 1078-1090, 1993
[8] R. Heidemann et al., " Demonstrating the enabling technology for broadband wireless local loop: ACTS FRANS and other hybrid fiber radio trials," Conf. Proc. ISSLS '98, Venice, Italy, pp. 401-407.
[9] J. J. O'Reilly et al., " Optical generation of very narrow linewidth millimeter wave signals," Electron. Lett., vol. 28, pp. 2309, 1992.
[10] M. Varasi et al., " Velocity matched modulators for wide bandwidth, high efficiency, and high extinction ratio," presented at the FOTONICA '99, Trento, Italy, June 2-4, 1999.
[11] H. Harada, H. J. Lee, S. Komaki, and N. Morinaga, " Performance analysis of fiber-optic millimiter-wave band radio subscriber loop," IEICE Trans. Electron., vol. E76-B, no 9, pp. 1128-1135, 1993.
[12] R. Sabella, "Performance Analysis of Wireless Broadband Systems employing optical fiber links," IEEE Trans. Commun.Vol. 47, no.5, May 1999.
[13] H. Mizuguti, T. Okuno, S. Komaki, N. Morinaga "Performance Analysis of fiber link for microcellular mobile communication system", IEICE Trans. Electron. Vol. E76-C, No 2, 271-278, 1993.
[14] O. K. Tonguz et al., " Personal Communications Access Networks Using Subcarrier Multiplexed Optical Links," J. Lightwave Techìnol.,Vol. 14, June 1996, pp. 1400-1409.
[15] T. S. Chu and M. J. Gans, " Fiber Optic Microcellular Radio," IEEE Trans. Veh. Technol., Vol. 40, No. 3, 1991, pp. 599-606.
[16] R. Sabella, M. Paciotti, and A. Di Fonzo, " Analysis of clipping effects on AM-VSB CATV subcarrier multiplexed optical systems," J. Opt. Commun., vol 19, pp. 72-74, Apr. 1998.

Roberto Sabella was born in Roma, Italy, in December 10th 1962. He received the degree in electronics engineering (Laurea in Ingegneria Elettronica) from the University "La Sapienza" of Roma, Italy, in 1987. He then joined Ericsson Telecomunicazioni, Italy, where he was a Hardware Designer for about one-and-a-half years, and then as a Researcher on advanced fiber-optic communications. His research interests are in high-speed optical communication systems and optical networks. Since 1991, he led a group working on the modeling and simulation of advanced communication systems. He holds two patents on optical cross-connects, is co-author of a book on high-speed optical communications and author of more than 70 papers on scientific/technical journals and international conferences. Since May 1997, he has been with CoRiTel consortium as research technical co-ordinator. Since 1999, he is the research manager of the Ericsson research unit in Roma.
He has been Lecturer (professore a contratto) at the University of Rome "Tor Vergata" and at the Polytechnic of Bari, Italy. Mr. Sabella is member of the IEEE/LEOS Technical Committee on Optical Networks and Systems, and member of the editorial board of the new international journal Photonic Network Communications (New York: Kluwer Academic). For the same journal he has operated as guest editor for the special issues on WDM transport networks. Moreover, he has been one of the guest editors of a special issue on optical networks, for the journal Computer Networks (Elsevier).
He is also a guest editor for the feature topic "Optical Networking Solutions for Next Generation Internet Networks", to appear on the IEEE Communications Magazines.

GPSR Compliance

The European Union's (EU) General Product Safety Regulation (GPSR) is a set of rules that requires consumer products to be safe and our obligations to ensure this.

If you have any concerns about our products, you can contact us on ProductSafety@springernature.com

In case Publisher is established outside the EU, the EU authorized representative is:

Springer Nature Customer Service Center GmbH
Europaplatz 3
69115 Heidelberg, Germany

Batch number: 10755292

Printed by Printforce, the Netherlands